AN INTRODUCTION TO
THE LOGIC OF THE SCIENCES

The physical sciences are complex and indeed marvellous intellectual constructions. They enable us not only to describe much of the physical world accurately, but also to explain many of the processes and kinds of things we find in it. To understand how the natural sciences achieve their power it is necessary to stand back from them and to study the principles of reasoning which underlie their use. Three main schools of philosophy of science have emerged. Logical empiricists have concentrated their attention on the logical structure of scientific discourse and the immediate descriptive and predictive power that the creation of that discourse gives human beings. Conventionalists have been more interested in the languages of science, and how these provide us with the ability to grasp and to classify natural phenomena. Realists have been more concerned with the power of scientific thought to go beyond experience to provide glimpses of the hidden powers and structures of natural things. This book is in the realist tradition, and its main aim is to show how science has the power to extend our understanding beyond that which we can know by observation. But it is also an attempt to incorporate in a larger scheme many of the insights of the empiricist and conventionalist positions. In this new edition a fourth force in the study of science, sociological analysis, is described and its strengths and limitations discussed. The book is intended for those who have some knowledge of or interest in the physical sciences, and would like to gain some understanding of how they have their astonishing power to reveal the natural order in the physical world. But studying science is also an aesthetic experience and, almost unique in works in the philosophy of science, a detailed treatment of the basis of scientific aesthetics is provided.

Rom Harré is Lecturer in Philosophy of Science at the University of Oxford and Adjunct Professor in Philosophy of Social and Behavioural Sciences at the State University of New York.

He was born in New Zealand and studied engineering and mathematics at Auckland University, lecturing in mathematics at the University of the Punjab, Lahore, Pakistan. He studied philosophy at Oxford, subsequently holding a Research Fellowship at the University of Birmingham. After lecturing in Philosophy of Science at Leicester University he returned to Oxford in 1960.

He has held visiting professorships at the Universities of Wisconsin; New York at Buffalo; Reno, Nevada; Aarhus; Kingston; Gothenburg; the Institute of Advanced Studies, Vienna and Padua; and has lectured at many universities around the world.

Other books by the same author include

THEORIES AND THINGS
MATTER AND METHOD
THE METHOD OF SCIENCE
THE PRINCIPLES OF SCIENTIFIC THINKING
THE EXPLANATION OF SOCIAL BEHAVIOUR *(with P. F. Secord)*
CAUSAL POWERS *(with E. H. Madden)*
THE RULES OF DISORDER *(with P. Marsh and E. Rosser)*
SOCIAL BEING
GREAT SCIENTIFIC EXPERIMENTS

Q
125
.H37
1983

AN INTRODUCTION TO
THE LOGIC
OF THE SCIENCES

ROM HARRÉ
University Lecturer in Philosophy of Science
Fellow of Linacre College, Oxford

Second Edition

WITHDRAWN

HIEBERT LIBRARY 59308
Fresno Pacific College - M. B. Seminary
Fresno, Calif 93702

© Horace Romano Harré 1960, 1983

All rights reserved. No part of this publication
may be reproduced or transmitted, in any
form or by any means, without permission

First Edition 1960
Reprinted 1963, 1965
Second Edition 1983

Published by
THE MACMILLAN PRESS LTD
London and Basingstoke
Companies and representatives
throughout the world

ISBN 0 333 34180 5

Printed in Hong Kong

Contents

PART ONE: SCIENTIFIC DISCOURSE

PART TWO: DISCOVERY AND CONFIRMATION

Preface to the Second Edition

This book was originally written for scientists who had some interest in the structure and method of the natural sciences. In its emphasis on the constructive use of models and analogies in scientific thinking it was in marked opposition to the logicist trends of the day. In the succeeding twenty-five years the idea that the main features of science as a cognitive enterprise could be understood through the logical structure of scientific discourse has fallen into general disrepute. In consequence little was needed in the chapters on scientific explanation to bring the argument up to date. However there has been a more radical development in the understanding of science, the realization that the historical situation of theories and empirical methods is an essential part of their force and power of conviction. In common with most innovations in philosophy this insight has been somewhat over-enthusiastically developed to the point that some of its proponents have been prepared to say that there is nothing to epistemology that cannot be explained in sociological terms. The importance of the debate that has developed around the sociological 'reduction' of the philosophy of science has led me to add a fairly substantial chapter, setting out the main lines of argument and modes of analysis offered by the sociologists of science. I hope both to have shown the importance of paying attention to the context of scientific work, and the impossibility of eliminating many of the traditional philosophical issues.

Preface to the First Edition

When student scientists are persuaded to read some logic and philosophy they are not easy to convince of the relevance of these studies to what they learn in lectures and laboratories. It has been my aim in this elementary textbook to try to make every point in as concrete a scientific setting as possible, and establish relevance by example rather than by precept. There are views about causality, about substance, about existential judgements, about the reality of theoretical entities implied throughout. I have tried to keep them just below the surface rather than discuss them explicitly, in the hope that the right questions will come to mind and be asked and answered in tutorials and seminars rather than expounded in lectures.

The plan of the book is dictated by two more particular aims which experience has taught. The method of teaching in science leads student scientists to suppose that everything they learn is on the same footing: roughly facts. They benefit greatly by having brought home to them the degree to which theory is concerned only remotely with fact, and more directly with models, their construction and exploitation. The second aim derives its importance from the same source, since it is also common to find that student scientists are unaware of the complexity and often subjective character of the criteria that are used in judging the rightness or wrongness, the satisfactoriness of theory.

I have also tried to bear in mind the needs of the increasing number of arts undergraduates and philosophy graduates who make their first acquaintance with the sciences through a book such as this. Too often in the past the intelligence of such people was insulted by confining the discussion to such untypical statements as 'All swans are white' with the consequence that the logic of single generalizations was widely mistaken for the whole logic of science, as if theory did not have any features which specially characterized its structure and the criteria for its assessment. The scientific examples have been chosen in such a way as to invite further reading by the arts student, in both biology and the physical sciences. The technical terms used have been chosen with the idea that they can be understood by reference to any reasonable scientific encyclopedia, or work of general science of a reasonable standard.

In this book will be found, I hope, a basis for philosophizing, not merely a set of principles.

Acknowledgments

I have to thank the Senate of Birmingham University for the opportunity to write this book. I would also like to express my thanks to Professor A. E. Duncan-Jones and Mr Bernard Mayo who gave me considerable help and advice; and to my wife who prepared the index. I am grateful to the editors of *Theoria* and *Philosophy* for their permission to use material which first appeared in their journals.

Part One
Scientific Discourse

1 Description and Explanation

One of the most marked features of the world as we know it is the never-ending succession of happenings. Most of our lives are spent either coping with the things that have happened or taking steps to initiate happenings on our own account. In the sciences we have systematic and highly sophisticated methods for managing and understanding happenings. A useful way of studying the methods of the sciences is to imagine them as means for answering certain kinds of questions about happenings. There are many kinds of questions we can ask about happenings and not all can be answered scientifically. We can give clear and precise answers to such questions as 'What happened?', 'How did it happen?', 'Why did it happen?'. However, it is very doubtful if clear and precise answers can always be given to questions like 'What was the purpose of the happening?', 'What was the point or reason of the happening' and 'What was the significance of the happening?'.

One sort of question for which generally satisfactory answers have never been found is that in which the purpose of a happening is demanded. It is not difficult to answer such a question when oneself or other people have been involved in the inception and effect of happenings, but it appears to be an intractably difficult problem to answer it in general and without reference to human interests. Many events can only be seen to have a purpose against a background of theology where, in the nature of the case, the scientific demand for confirmation can never be satisfied. Even if such confirmation were forthcoming it would be difficult to make one's general statements of purposes clear and precise. This is simply because what is taken to be the purpose of a happening depends very much on the character and history of the person concerned. This difficulty arises with other questions of the same kind. If we ask for the point or the significance or the reason of a happening the answers we give depend very much on who we are.

Other questions in the list such as 'What happened?' and 'How did it happen?' are certainly not easy to answer precisely and clearly but the difficulties involved in giving answers are technical difficulties not logical ones. The invention of instruments and the use of a precise, technical vocabulary enables people answering these questions to eliminate much of their own bias. When answers to such questions are given with the widest possible generality and the greatest possible precision, that is science. To 'What?' questions scientists answer with generalized and precise descriptions of the world; to 'Why?' and 'How?' questions with similarly generalized and precise explanations.

If we wish to understand the special sort of describing and explaining that scientists undertake we must first understand the way we ordinarily describe and explain the events that happen and the things that are found in the world. Describing and explaining are essentially linguistic activities, and though a picture can take the place of a description and a diagram of an explanation, both can be misunderstood in a way that descriptions and explanations formulated in a language cannot be. Before we know how pictures and diagrams are to be applied and how they are to be understood we need to know to which of their many features we should attend. This is something not shown on the face of the picture or the diagram. Words have a narrower range of ambiguity than pictures and it is in words, or in a conventionalized symbolism if we want greater exactness, that our most sophisticated and careful explanations and descriptions are expressed. A study of the workings of science must begin with a study of the language of description and explanation. We must begin with the logically simplest kinds of descriptions and explanations—those we formulate in everyday language to deal with everyday situations. In a way the study of the logical gradation from everyday to technical language parallels a study of the historical development of science, for it was out of untechnical descriptions and explanations that the technical languages of the sciences grew. This growth was stimulated by the need to express the results of discoveries that were outside our common experience and the need to satisfy the increasing demands for rigorous presentation that reflected the growing desire of scientists for greater exactness and certainty and for a wider range of expression than every-day language was capable of providing.

DESCRIPTION

Descriptions must, for the most part, be given verbally. In order to make an analysis of descriptions we may need to analyse very complex linguistic structures so that we must find, first of all, the most convenient units of language. A printed page provides us with two ready-made units, the word and the sentence. It is sentences which we use to convey the simplest descriptions, for they are, as it were, the atoms of language. The differences between them are to be ascribed to their internal structure and the words that make them up, as in physics we account for the differences between atoms by their internal structure and the differences in arrangement and character of the minuter 'particles' of which they are built. In studying descriptions we shall begin by analysing the simplest descriptive sentences, in order to find out how they work and what particular jobs the words within them do.

One of the things we want to know about the descriptions we give and particularly those that other people give is whether they are true or false. We need to be clear at the beginning of our study of language which particular linguistic units, utterances, sentences, statements or what, are the proper subjects of judgements of truth or falsity. The usual custom[1] is to make a distinction between sentences and the use of these sentences on particular occasions to make statements, issue utterances and make assertions. A sentence, a certain form of words, is thought of as a tool which can be used over and over again to make many statements, as a hammer can be used over and over again in the hammering of many nails. Just as the hammering can be successful or unsuccessful with the same good hammer, so statements made with the same sentence may be true on one occasion and false on another. A statement will be true if the sentence is used appropriately and false if it is used inappropriately. So though our analysis will be carried out on sentences (the dead bodies of language), our judgements of truth and falsity must be made of statements. They must be made when sentences are brought to life by being used in some definite context to make serious assertions, assertions which we would stand by until they should be proved wrong.

There have been many analyses proposed for the simple descriptive sentence, the sentence which can be used to ascribe a property to some definite thing. The most useful for our purpose is that of G. Frege[2]. Frege's analysis depends on the recognition that a descriptive statement does two jobs; it individuates and refers to a certain subject, and it ascribes to this subject a certain property. The first task of the analyst is to isolate the expression or expressions which carry the primary reference. This Frege called the *argument* of the sentence, the remaining part the *function*. Consider a very simple example, where the reference is carried by the simplest of all referring expressions, a proper name. If we remove the referring expression from 'Rutherford was a tall man' what is left is the sentence frame '—was a tall man'. The gap can be filled with all sorts of referring expressions, 'Caesar', 'Faraday' and so on, and new sentences generated, each of which can be used to make assertions that may be true or false. Frege pointed out that here was an analogy with mathematical functions. If we look upon x in $x + 3 = 6$ as marking a gap where definite numbers can be put then substitution of 3 for x gives a true statement while substitution of any other number gives us a false statement.

We need some additional elements for a complete analysis of simple descriptive language since the above analysis applies only to those sentences which are logically 'atomic', that is which cannot be analysed into complexes of simpler function-argument units. The additional elements are sentence-connectives. What matters with simple descriptions is whether they are true or false and the kind of sentence-connectives that we need will reflect this, for they are conjunctions with specific truth-properties. The simple conjunction 'and' is a good example. In many cases when it is used to connect sentences which are used to make true statements, the whole complex is itself a true statement. There is a definite answer to the question 'Is this complex assertion formed by the use of "and" true or false?' for each combination of truth and falsehood among the simple statements that make up the assertion. Suppose we link 'Jack went up the hill', a true statement, with 'Jill went down the hill', a false statement. The compound statement 'Jack went up the hill *and* Jill went down the hill' must be false. Similarly two false statements when compounded by 'and' make a compound false statement. There are many conjunctions which have truth-properties like 'and'. For example if statements are linked by 'or', 'not both . . . and . . .', 'if and only if' and many others the truth or falsity of the compound statement can be worked out from the truths and falsities of the atomic statements used to make it up. Compound statements of this sort are called *truth-functional*, and an analysis of language on this basis a truth-functional analysis. It was thought at one time that all the statements made in the sciences could be analysed into such compounds of atomic statements. All we had to do to be able to work out the truth of any scientific statement, however complex, was to find out the truth or falsity of the atomic statements that made it up. When we have seen, in the course of this book, just what a great diversity of kinds of statement there is in the sciences, we shall see that though such a view may be appropriate to very simple descriptions of the kind we have been discussing it is inadequate as a general account of science.

One of the questions we shall have to ask ourselves about the sciences is how we tell whether to accept or reject a statement as an adequate description. This question is a very complex one and not easy to answer for the many kinds of statement that appear in the sciences, but it can be answered fairly easily for the very simple descriptive statements we are analysing in this chapter.

A true description is a successful attempt at describing something. What are the criteria for this success? To answer this question we must first ask what the process of describing involves. This is not meant to be a psychological question, but a logical one. It is to be answered by an investigation of the conditions under which we would call a sentence descriptive. Logically speaking the process of description involves two steps; the selection of something to be the subject of the description, and the recognition of the thing selected as belonging to a certain kind, or as possessing a certain property. The expression in a suitable linguistic form of this selection and recognition is a description. The two steps could be thought of in another way. We might think of ourselves identifying the thing to be described and then matching it against a sample, pattern or type to see what sort of thing it is. The process of description could be thought of as the recognition that a certain thing is, as C. S. Peirce called it[3], a token of a certain type; that is that a thing is a member of a certain class or a representative bearer of a certain property. An example of this mechanical process of describing may be helpful in bringing us to see what are the basic logical requirements of a successful description.

Suppose we have a set of objects, each identified by having a symbol *N*, *M*, etc. marked upon it. Suppose further that we have an interest in, say, turtles. We can ask two questions about N: (i) we can consider N just as an individual thing which we want to describe; (ii) we can also treat N as an example of a class or kind or species, and ask questions about it in this aspect too. These questions are:

I (1). Which object here is a turtle?
Ans. N.
I (2). What is N?
Ans. A turtle.
II (1). Which of these objects can be used as an example of the turtle?
Ans. N.
II (2). What is N an example of?
Ans. The turtle.

We could imagine a mechanical procedure for giving these answers. Suppose the recognition of a thing as being of a certain kind is dependent upon matching it against a card bearing a picture of a representative member of the class. Suppose further that we have a set of such cards each with a different animal pictured on it. To find out which object in our set is a turtle we can select the turtle card and match the objects against it until we find one, say N, which is a proper match. We can from this discovery answer questions I(1) and II(1), by reading off from the object its individuating symbol. We can answer the latter as well as the former because once we have been satisfied of the match we could use N instead of the turtle-card in further identifications. To find out what sort of thing N is, and so to answer questions I(2) and II(2), we could try N against various cards until we found a match.

Seemingly what we require in order to give a description are verbal analogues of pointing, identifying and matching. In my example the functions of pointing and identifying (in general of referring) are carried out by the use of an identifying code, the capital italic letters. Once the correlation of the objects with the code signs has been worked out there is no further need for pointing in giving a description. Pointing can be confined to the prior process of correlating the code with things as we correlate names with persons in baptisms and introductions. The verbal analogue of matching an object with a card is the use of appropriate predicates to describe it. One can imagine in a very primitive situation all three elements—object, word and sample (card)—being required in the process of giving a description, but in every-day life recognizing that a thing is of such and such a kind and using property and class words correctly of it are not distinguishable activities. The criteria for the success of an elementary descriptive statement are very simple:

(i) Reference to the object we intend should be ensured by the use of the referring expression, such as a name, which is conventionally correlated with it.

(ii) The object should really have the property or belong to the class that we say it does and we should use the property or class word in our descriptive predicate that is conventionally associated with a standard instance of this property or a sample member of the class.

The use of uniquely referring expressions like names is, however, comparatively rare and only of practical utility when the number of objects to which we want to make reference is small (hence the use of names for the people we know). In a great many cases the reference of a sentence is carried by a descriptive phrase of such a kind that only one object can be its referent. *e.g.* 'the tallest man alive'. Such a phrase has come to be called a *definite description*. In other cases the uniqueness of reference is left open as when we talk of 'a tall man'. To use such expressions as these we need two samples in every act of description; one to identify the object from the descriptive referring expression and one to enable us to determine the correct property to assign to it. If we analyse 'The last apple in the box is rotten' we get as a referring expression 'the last apple in the box' and the ascriptive expression '—is rotten'. In terms of our card analogy we need a card for recognizing the last apple in the box as well as a card for recognizing rottenness. In linguistic terms we need to be able to say 'This is the last apple in the box' and 'This (the last apple in the box) is a rotten apple'.

I have so far used 'having a certain property' and 'being a member of a certain class' as if these two expressions were equivalent. Their close connection is expressed in the classical philosophical distinction between *intension* and *extension*. A class or group of things isn't just a haphazard collection but a grouping of things according to some principle of selection. We generally put things into classes when they have one or more properties in common. For example people are said to belong to the same social class when they have certain habits and customs in common.

These habits and customs serve to differentiate them from people in another social class. Those properties which all things in a class have in common and which serve as the criteria of selection for membership of the class are called the class *intension*. The things which are selected in this way are called the class *extension*. My discussion up to this point has explicitly been concerned with particular and individual things, but we have had to make reference to classes whenever any linguistic activity beyond simple naming has been under discussion. This is because many things may have the same property or properties and so be correctly given the same description. Indeed the recognition of the possibility of the same descriptive predicate serving to describe many different things underlies the basic analysis of sentences that we have been studying. The possession of a certain property (the intension of a class) distinguishes members (the extension) of a class from non-members. There should be a way then of using the elements we have found to be necessary for a definite description to refer not to one member of a class but to all members or to the class as a whole. A statement which says something about all members of a class or a class as a whole is called a *generalization*.

I. *Member generalizations*. The process of applying a certain predicate to all members of a class having a certain defining property is generalization. It is the conveying in one statement of the information that would require many statements of particular fact to convey bit by bit. It is instructive to study generalization from a grammatical point of view. The word 'all' can be used with an intensional expression (usually the predicates of a definite description) to refer to the whole class defined by that expression. For example a singular descriptive statement 'My cat likes cream' can be generalized to refer to the class of cats by the use of the world 'all'; 'All cats like cream'.

In many cases the predicates of the definite description of an individual member of a class are condensed into a single class-expression. For example instead of referring to the animal with the long nose, large ears and grey skin, etc. we can use the word 'elephant' as a class-expression. The definition of this word will be the expansion of the intension of the class into the predicates that would serve for a unique definite description of any elephant. We are now in possession of a neater generalizing device than that above, for now we use the single word for the class where we had to express the intension of the class in a set of predicates. There are a number of grammatical alternatives for generalizing with a class-expression each having its own proper use. In addition to 'all elephants' we have 'each elephant', 'every elephant', 'any elephant' and so on. These devices emphasize the fact that the range or class to which we refer is made up of individual members. It is also possible with expressions like 'elephant' to make reference to the whole class as if it were a kind of extended individual. In 'The elephant is indigenous to the forests of Central Africa' each single elephant is of significance only in so far as it forms part of the complex, extended 'individual', the elephant-species.

Not all assertions in generalized form genuinely convey generalized information, facts about classes of individuals. Class-expressions such as 'elephant' which serve to condense groups of predicates have to be defined. In practice this is almost always by ostension, that is by showing the learner an elephant. It must always be possible though to say what it is about the animal that distinguishes it from, for example, a mammoth. Statements of this kind, which bring out the intension of a class by expanding the class-name, are usually expressed in the generalization form and are apparently no different from those generalizations which genuinely convey information. A statement like 'All elephants are grey' can be treated as obliquely expressing a rule for the use of the word 'elephant'; for instance 'Only use that word for animals which are grey'. Whether this is an appropriate way to read the statement can be gleaned from the response of the linguistic community to the appearance of a 'doubtful case', such as a large, long-nosed mammal which is white. Qualifying the statement to 'Most elephants are grey' demonstrates the propriety of an empirical reading, while qualifying the dubious animal—'That's not an elephant, it isn't grey' expresses a definitional reading. Statements of which this is the case are called *analytic statements*. In contrast to these are *synthetic statements* whose truth is determined by inspection of the world.

We should notice in passing that there is another class of general analytic statements whose truth depends upon the meaning of words. The 'elephant' example depended upon the meaning of the word 'elephant', a word which could, and often is, used to convey information. Another class of analytic statements depends on the meanings of the words which are used, not to convey information, but to build up the structure of the sentence used to make the statement. These are the *tautologies*. For example 'Either it is an elephant or it is not an elephant' depends for its truth, not on the meaning of 'elephant', but on the meaning of the structural words 'either', 'or' and 'not'. It is not always clear whether a statement is genuinely informative or whether it is a disguised tautology. In general it is important to distinguish those statements which are analytic assertions and convey no factual information about things from those which are synthetic and so can convey facts about the world.

The strict distinction between analytic and synthetic statements has been queried in recent years, largely on the grounds that the semantic rules upon which some at least of the former are based are contingent features of the language. Quine, for example, holds that at least some analytic statements are synthetic linguistic generalizations. More telling has been the feeling that a sharp distinction between linguistic generalizations and generalizations about non-linguistic matters is somewhat artificial. For instance non-tautological 'analytic' statements depend on matters of fact, such as the pervasive greyness of elephants, which are taken up, as it were, into the elephant concept. It makes sense to make a property of members of a class of things a criterion for using the class word (and so part of its meaning) only if we have reason to think that all things of that kind manifest the property, and if any do not, there is some accidental reason why their tendency to manifest it has been aborted in that case. (Of course these moves require that there be some other property or properties which also tentatively mark off a species or kind.) Nevertheless, for the logic of science, it is still worth emphasizing the distinction between displaying the general features of concepts and displaying the general features of things or substances.

II. *Substance generalizations*. Naming and sample-matching is not the only way of describing the world, nor is member generalization, depending as it does upon the notions of an individual thing and a collection of individuals, the only way of conveying generalized information. Another method of description begins with what has been called *feature-placing*[4]. We can recognize and talk about substances as well as circumscribed individuals. Substances are those stuffs which form large scale features of the world. Their recognition has always been important in science. Early scientific investigators recognized four main substances, earth, air, fire and water. The logical requirements for saying something about substances are similar to those we have already noticed as being necessary for saying something about individuals. There should be a referring device to locate what we are saying in some point in space and time and a predicate by means of which we ascribe something to our referent. If we imagine that our descriptive language consists largely of substance words then a description will be given by pointing to something and saying 'This is such and such a substance'. In expressions like 'This is gold', 'Here is water' there is no grammatical subject and hence no individual referring expression for there is no determinate object to which reference can be made. We want to say that the part of the world to which we point exhibits or is the feature in question.

This kind of descriptive expression is logically so primitive that it can be called neither particular nor general. Particularity can be gained by circumscribing a portion of the feature to form a pseudo-individual to which individual reference can be made. For example the expression 'stretch of' with an appropriate feature word becomes an individual referring expression and can be used as a definite description in assertions like 'This stretch of forest is surprisingly green'. Words like 'plot', 'part', 'slice', 'bit', 'piece' and 'portion' are used to delimit features with more or less exactness and so can be used for the building up of individual referring expressions. On the other hand generality can be gained by using the feature word as the subject of a sentence, in much the same way as class words (as condensed descriptions) are used. For example substances like water can be referred to in a general way by the use of the feature word in assertions such as 'Water is an excellent solvent'. Non-informative generalizations serving to state the defining properties of substances are found with feature words in a similar way to their appearance with class words. Similar too is the difficulty that often arises as to whether in fact the generalization is a definition— a difficulty which can only be solved by careful attention to context.

I have tried to show the grammar of generalizations as a natural extension of the grammar of individual description. It is in point now to sum up what has been said and to initiate a new step by asking what it is that generalizations do for us and to what uses we can put them. Economy of thought is an obvious aim in description and generalization clearly satisfies it without loss of adequacy, for in a general statement the least that we do is to sum up the information conveyed by many particular statements. I think it would also be granted that when we are in possession of a generalization we have a greater grasp of the facts conveyed by it than we could have if we were confined to numerous particulars. In fact part of what we mean by gaining a grasp or understanding of something is coming to see it in a class or group, as an instance of a generalization. Something cannot be seen as an instance of a generalization unless that generalization is already held and understood. There is still another job which generalizations do for us. They play an important part in reasoning for our rules of inference are derived from them.

Suppose one were asked to judge the validity of some simple argument, say:

Whales are mammals
therefore
Whales are warm-blooded

The simplest way would clearly be to consider the conditional statement 'If a species is mammalian then it is warm-blooded'. If this is true then the argument is sound or valid; if it is not true then the argument won't do. It is, as we say, unsound or invalid. It is sometimes said that conditional statements like 'If a species is mammalian then it is warm-blooded' are rules according to which valid arguments are constructed. More picturesquely it is sometimes said that conditionals are licences which permit one to make certain inferences. An argument is valid (an inference proper) when the conditional statement which governs or licences it is true. From the *antecedent*, the clause following 'if', we may infer the *consequent*, the clause following 'then'. Similarly it it were not true that if a species is mammalian then it is warm-blooded an argument constructed in accordance with it would have to be judged unsound; and an inference licensed by it improper. The consequent would not be a sound inference from the antecedent. In the end then judging an argument can be reduced to judging the conditional statement upon which its validity depends. But how are conditional statements to be judged?

Conditionals have a hypothetical air, and when one starts talking about what happens if so and so is the case one might seem to be loosing a little of that hold upon hard facts which is supposed to be such a feature and such a virtue of the sciences. However this hypothetical air is not to be taken too seriously for with every indicative conditional statement goes a corresponding general statement which matches the conditional truth for truth, falsity for falsity. If it weren't true that 'All mammalian species are warm-blooded' then it wouldn't be true that 'If a species is mammalian then it is warm-blooded'. It is generalizations that are at the back of inferences and arguments, being brought to bear through their corresponding conditionals.

Sometimes arguments are based directly upon generalizations. When one identifies a certain thing as being a member of a group or class then this identification depends on the thing having the defining properties of the class. In a general statement everything in the class so defined is said to have some further property. It follows then that when a thing has been identified as a member of a class we may also ascribe to it the further property which our generalization ascribes to members of the class. We don't need to bring in any conditional statement but simply go directly from the class to an instance of it. This is argument by *instantiation*. All that we require for argument by instantiation is that the instance should be less general than the subject of the generalization. For instance from the general statement 'Anthropoids are omnivorous' it follows that the same may be said of any species of this genus. And from '*Homo sapiens* is omnivorous' it follows that Charlie Smith, an instance of *Homo sapiens*, is omnivorous too.

There are many kinds of generalizations recognized by logicians some of which cannot be used for argument by instantiation. However, the subtleties involved in distinguishing these kinds would be out of place here. There is one subtlety which is very important for the logic of science and which cannot be omitted, that of the subjunctive conditional. The topic, however, is fairly technical and the next ten paragraphs might well be omitted at a first reading. I place this discussion here because the puzzles which arise from the use of the subjunctive mood in conditionals are not confined to the language of science but are quite general. Indeed some scientists and logicians have tried to find ways of eliminating subjunctive conditionals from the sciences altogether, so that puzzles originating in ordinary language should not be imported, by the use of that language, into science.

Indirectly a generalization lends support to a corresponding conditional, so that the acceptability of a certain conclusion inferred in accordance with the conditional may depend in the last analysis on the acceptability of a certain generalization. It would seem on first sight that there is a quite straightforward connection between general conditionals and generalizations. The acceptability of the two is so indissolubly linked that the one may be said to be an alternative form of the other. 'If a species is mammalian then it is warm-blooded' seems to be an alternative form of 'All mammalian species are warm-blooded'. The acceptability of the one involves the acceptability of the other. However, much reasoning is carried on not with indicative conditionals as I have described above, but with those employing the subjunctive mood in various ways.

In most of the examples that are of interest to us as students of scientific discourse, both the indicative and the subjunctive conditional are general, concerning species, types or open classes. The lawlikeness or natural necessity of a generalization seems to be manifested (at least in one way) by the fact that we are ready to accept the corresponding subjunctive conditional. Generalizations, one normally expects, have instances otherwise there would seem to be no point in making them; yet it is a feature of general subjunctive conditionals that they are put in such a way as to imply that, in some cases, the corresponding generalization has no instances. For example 'If protoplasm had developed in a much hotter environment then it would have been based on silicon rather than carbon' uses the subjunctive in such a way as to imply that there is no silicon based protoplasm. How, it may be asked, do we know that the general statement has anything to recommend it if it is not instantiated? Similar puzzles about the grounds of subjective conditional statements and their relation to categorical indicative statements arise where the reference of the statements are singular. 'If Caesar had not crossed the Rubicon he would not have been assassinated' carries the *implication* that he did cross and was assassinated, yet the content of the subjunctive statement seems to be a living Caesar north of the famous river. Subjunctive conditionals which imply that something didn't happen which we know very well did have been called *counterfactual conditionals*. A considerable literature[5] has grown up in the attempt to elucidate their logic. Not only are subjunctive conditionals important because of their intimate connection with laws of nature but they also seem to be involved in giving the meaning of *dispositional* properties. If we analyse a statement like 'Polythene is waterproof' we must include the statement that if a polythene container were to be filled with water it would not leak. To say of sugar that it is soluble is to say both that if it were to be put in water it would dissolve and that if any specimen had at any time been placed in water it would have dissolved.

There are two main problems: How does the fact that counterfactual cases seem to be involved in laws of nature affect our notion of their truth? How can we justify those general subjunctive conditionals whose associated general indicative statements, unlike our laws of nature, have no instances or near instances at all?

The problem of confirming the generalization which seems to be needed to justify the counterfactual conditional has greatly exercised logicians. It can be asked:

(i) How can the indicative general statement that backs up the conditional ever properly be confirmed since the use of the subjunctive in the conditional implies that the state of affairs described in the antecedent did not or has not occurred? Apparently this state of affairs is one in which we are interested for otherwise what would be the point of framing the conditional in the first place?

(ii) We may ask in what way the truth of any conditional depends upon the truth of its components, antecedent and consequent. This form of the problem has been discussed by those logicians who wished to find a truth-functional analysis for all kinds of statements[6]. Clearly we do accept counterfactual statements and even maintain that they are true knowing full well that their antecedents are unfulfilled and therefore strictly false. We may agree that if he had moved she would have shot him when we know in fact that he didn't move. It may be the business of a court to decide upon the truth of just such a counterfactual. It seems then that the truth of counterfactual conditionals does not depend in any direct way upon the truth of their components. Now these statements are quite commonplace and used unhesitatingly in ordinary discussions. There must be something astray with our statement of the problems they present if a common way of talking takes on an air of insoluble paradox.

A partial solution to these problems can be given by pointing out that our willingness to accept a general statement or categorical is not based only on the instances in which it is actually applied, unlike the grounding of singular descriptive statements. We do not commit ourselves to the same degree to being able to prove a general statement, by instances. Furthermore the determination of the acceptability of a conditional, and of the generalization upon which in the last analysis it depends, is a much less strict affair that the determination of the acceptability of a particular description. There is a whole set of terms by which we hedge round the acceptability of generalizations ('on the whole', 'in general', 'for the most part') which have no application to particulars, and which must infect the related conditionals when we have to depend only on instances. Some of the difficulties that are felt about conditionals and that find expression in the apparent paradox of true counterfactuals having false components can be removed by looking more closely at the means we actually use to support generalizations. Roughly a generalization is acceptable when it has been arrived at in an acceptable way from soundly grounded facts and has tended to be confirmed by subsequent experience, experience has thrown up no contrary existence, *and there is some theoretical reason for expecting it to hold*. Similar considerations apply to the corresponding and derived conditionals, general and particular. If they can be constituted as *possible* applications of the appropriate theory unfulfilled cases do not count against the reliability of the generalization. They are used to present hypothetical situations in which we might apply the conditional, not situations in which the conditional was applied and failed.

General subjunctive conditionals invite us to envisage a world completely different in some important respect from that with which we are familiar. They must, therefore, be comparatively rare in scientific discourse, for what they suggest is contrary to fact is not something accidental. The generalizations that correspond to them are very fundamental, stating laws of nature essential to our sort of world. For instance 'If there were no gravitational fields nebulae would not develop into systems of stars' invites us to envisage a universe different in a fundamental respect from the one we know. On the other hand the hypothetical revisions of fact required by particular subjunctive conditionals are not as deeply opposed to well-known laws of nature. The field of possibility sketched out by the contrary-to-fact antecedent in a particular case is usually generally acceptable to us as a reasonable alternative to what actually happened, or as a reasonable supposition as to what will happen. In recent years the theory of subjunctive conditionals sketched here (written in 1958) has been expounded as 'possible worlds semantics' by S. Kripke and D. Lewis[7]. Once the stage is set by the antecedent, the conditional, whatever its mood, functions as a simple rule of inference within the field of possibilities delimited by its antecedent. Much is condensed in a counterfactual conditional. 'If Hitler had invaded Britain in 1940 he would have won the war' can be analysed as follows:

(i) Field of Possibility. The invasion of Britain is undertaken by Hitler.

(ii) Rule of Inference. If Hitler invades Britain he will win the war.

(iii) Licensed Inference. Hitler wins the war, in the circumstances envisaged in (i).

The grounds for the rule of inference (ii) are the general truths of military logistics applied within the field of possibilities delimited by the subjunctive antecedent. This informal analysis can be made much more precise in terms of what are called *modal* statements[8], that is statements of possibilities.

With the help of our informal analysis we can now see that the role of a subjunctive conditional can be broken down into parts.

(i) The antecedent delimits a certain universe of discourse within which we are now to make inferences. It brings us to consider certain possibilities.

(ii) Within this universe the corresponding conditional in the indicative mood operates straightforwardly as a rule of inference. It gets its sanction from the general laws which operate in the field in which our possibility is envisaged.

(iii) In the case of particular conditionals conclusions are drawn within the framework of the known laws but not within the framework of the known facts. When we wish to state these conclusions alongside what actually happened they must be stated in such a way that they can be distinguished from the things which really happened. This distinction is achieved by making them grammatically subjunctive and logically modal. They are, as we say, among the possibilities of the situation (*e.g.* in 1940).

Subjunctive conditionals also have some importance in the analysis of dispositional predicates. These predicates are in fairly common use for both things and people, *e.g.* 'brittle', 'irascible'.

These are not simple predicates used for ascribing easily distinguishable and openly displayed characteristics. I may tell you that an object is brittle without smashing it in front of your eyes or that a now placid old gentleman is irascible without provoking him. 'Brittle', 'irascible' and others of the same family must include in their analyses statements of the form 'So and so would happen if . . .'; that is subjunctive, conditional elements. These elements are involved because our ascription of brittleness to a thing invites the consideration of certain possibilities, *e.g.* the thing suffering a sharp blow, and tells us what the consequences would be if any of these possibilities were realized, *e.g.* it would shatter. The justification of the conditional element in the analysis of dispositional predicates may be more or less general. It is less general when the grounds for ascription are the previous experiences we have had of the realized possibilities in similar cases, more general when the consequences of realized possibilities can be deduced from theory. The ascription of the disposition in these circumstances is no longer dependent on our witnessing some destructive accident. We may be able to predict, using our knowledge of the structure and origin of a certain material that it will be brittle without our having to make and smash specimen artefacts.

Let us turn again now to a more general consideration of the relationships that hold among conditionals and between them and the general statements from which they are derived. These relationships are important for our discussion of the role played by generalizations in science, for just as in ordinary language, they lie behind many pieces of scientific reasoning. Provided that the indicative mood is preserved in the change from categorical to conditional form there seem to be no difficulties. Information in one form is transformed into information in another form. The difference is grammatical for there is no change of content. This applies both to general and particular conditional statements. The relations that exist between the various forms and quantities can be expressed in a diagram.

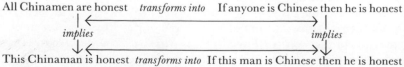

Conversion of the conditionals to the subjunctive mood alters this network of relations. For the general indicative forms we could equally well have written 'justifies' for 'transforms into'. However the single indicative generalization above is not sufficient to justify the general subjunctive conditional 'If anyone were a Chinaman he would be honest', still less will it serve to justify 'If there were Chinamen they would be honest'. In the latter case by our use of the subjunctive mood we

(i) imply that there are now no Chinamen,
(ii) delimit a universe of discourse in which such people may be envisaged. Since we have no generalizations about these people their characteristics must be derived from some general laws of nature by means of which we can predict how people of a certain kind, if they existed, would behave.

In the diagram above it can be seen that the singular indicative conditional 'If this man is Chinese then he is honest', to be justified in default of particular information about this Chinaman, must be referred to the generalization 'All Chinamen are honest'. A singular indicative conditional says something about an instance of its corresponding generalization, while a singular subjunctive conditional says something about a possible instance of its corresponding generalization. A singular subjunctive conditional can be justified only by reference to a statement more general than any into which it can itself be transformed. Similarly a general subjunctive conditional is not justified by its corresponding generalization, but rather by reference to generalizations more general than any with which it corresponds. When we delimit a possible situation by means of a subjunctive antecedent in a conditional we can justify this only if it can be shown that the situation is possible. This can only be by reference to what we know about such situations in general. We can feel justified in holding an indicative general conditional on the grounds of its being an alternative form of a generalization that is grounded in the facts. When possibilities are in question there are no supporting facts for a corresponding generalization. Rather there are other generalizations which give it support. We can make appeal only to more general information in determining the reasonableness of a possibility; in designing a plausible (so possible) world.

EXPLANATION

We ask the question 'Why?' mostly about happenings. Since it is in reply to this question that explanations are commonly given it is happenings that we usually want to explain. To give an explanation is to give the reasons for a happening. There are all kinds of explanations since there are a great many different considerations that can count as reasons. The logical character of the various kinds of explanations depends upon the sort of happenings that we are asked to explain, and the sort of reasons we are satisfied to accept by way of explanation. If the preliminary analysis of discourse undertaken in this chapter were intended to develop either into a study of the language of morals or the language of psychology the interesting cases of giving an explanation would be derived from studying the ways we give reasons for our own and other people's actions. As this preliminary analysis is aimed at elucidating the anticipations in everyday language of scientific discourse it is to happenings that are not connected with human aims and wishes that we turn. Scientific explanation aims at giving the reasons for happenings that are independent of human caprice. When human action itself comes under a scientific scrutiny this is commonly thought to be scientific in so far as it treats human actions as happenings on the same footing as happenings unconnected with our aims, purposes and wishes.

Nevertheless as a preliminary to the understanding scientific explanation the logic of ordinary, everyday explanation must be explored. The use of words like 'reason' includes both accounting for natural happenings and for what people do. In order then to carry out the programme of the last paragraph we must first ask what is involved in giving a reason in general, and we can then see how the concept can be made to apply to happenings independent of people. When we are asked to account for our actions, to give reasons for them, say by a policeman, then what we do is to list our intentions and the features of the situation that led to these actions. However, for a statement of a feature of the antecedent situation to be accepted as a reason for the action it must also be apparent that it is relevant to the action. A completely specific account must include statements which make clear the relevance of the feature of the antecedent situation to the action, as the realization of our intentions. So much can be relevant to an action that in practice a more circumscribed account of the antecedents to the action must be given. In the natural science case where there are no intentions involved reasons become causes. It has come to be generally agreed that to ask for the cause of a happening is to ask for the antecedent conditions which are both necessary for it to occur and sufficient to bring it about. But this is to ask for all the relevant features of the antecedent situation as the account of the happening. In practice the cause is usually picked out from the plethora of necessary and sufficient conditions by identifying it with that which by changing brings about the happening in question. Stable background conditions, though necessary, are not usually cited as causes[9]. An explanation of a happening involves the statement of the cause of what happened.

An explanation of a particular happening has then the following features:

(i) It will give a reason for the happening by mentioning a certain feature or features of the antecedent situation.

(ii) It will either imply or state directly the relevance of the feature or features in question to the happenings for which an explanation is wanted. It is from the different ways in which these requirements are satisfied that the characteristic features of different kinds of simple explanation spring.

I. *Linear Explanation.* When we answer a query as to why a certain happening took place with a statement of another particular happening we are giving what I shall call a *linear* explanation. The word 'linear' is intended to emphasize that this kind of explanation is given in a statement of the same logical status as the status of what is explained. A particular happening is given as an explanation of another particular happening. Linear explanations are very common. For example when we ask 'Why has the fire died down?' and are given such an answer as 'The flue has become blocked' we are given one particular fact as an explanation of another. But why, it might be asked, is 'The flue has become blocked' acceptable as an explanation rather than any other particular fact in the antecedent situation, say 'I have just folded my arms'? What belief would be necessary to lead us to accept the latter as an explanation? Clearly the belief that a connection exists between armfolding and the dying down of fires. The reason that the statement of the particular happening, the blocking of the flue, is acceptable as an explanation of the dying down of the fire without further comment is that the connection between fire, flue and draught is such a commonplace fact that its explicit statement can be dispensed with. The acceptance of a certain particular happening as the basis for an explanation depends upon the connection of this happening and that which is to be explained being just such a commonplace. That this is the case can be seen from the way in which a linear explanation might be challenged. One disputes not so much the particular antecedent condition put forward in the linear explanation, but its relevance to the happening. To dispute the relevance of the condition is to dispute the connection between cause and effect in a general way. The rejection of the flue explanation, supposing it to be true that the flue is in fact blocked, could be made cogent only by a wholesale rejection of all such explanations, that is by a rejection of the generalization stating the general connection between fires and draughts. We might say 'Draught is not important, it's having plenty of fuel that counts' and so the relevance of the draught explanation in general is rejected and a general alternative proposed.

II. *Hyperbolic Explanation*. In the linear explanation the particular happening taken as the cause is stated and the general connection of this happening with the happening to be explained is understood. In what I shall call *hyperbolic* explanation the general connection of an antecedent to the happening to be explained is given and the particular happening that is the cause is understood. For example to 'Why is the fire dying out?' the reply 'Fires die out when their flues get blocked' is a hyperbolic explanation. I use the word 'hyperbolic' to suggest that there is a difference in logical status between the explanation and what is to be explained. The hyperbolic explanation is general while what is to be explained is the fate, in our example, of a particular fire. However, the adequacy of the explanation does not seem to be symmetrical with the adequacy of the linear explanation, for their seems to be a strong tendency to make the application of the generalization specific and explicit by stating in the particular that *this* is the flue that is blocked. The explanation of this asymmetry lies I think in the way we direct ourselves in looking for an explanation. The direction in which we look for a cause, a relevant antecedent happening, is determined to a large extent by our prior knowledge of what is relevant to a certain kind of happening. It follows that logically prior to our giving a linear explanation is the acceptance of a certain generalization, an acceptance of certain relevancies. When we do select a cause and state it as the explanation of another happening we have, as it were, completed the logical requirements for explanation in both giving the antecedent happening and determining its relevance. In giving a hyperbolic explanation we stop the explicit process of giving an explanation half-way; leaving the inference to what must be understood until after the explanation has been given. In giving a hyperbolic explanation we are laying down the direction in which we must look to find the cause, leaving the actual selection among the antecedent happenings to be understood. The asymmetry exists because we cannot say what is the relevant particular until the generalization has been stated, or is understood. In a linear explanation what is understood is prior to the explicit explanation, in a hyperbolic explanation consequent to it.

III. *Explanation in Detail.* What we might call an *explanation in detail* combines the linear and hyperbolic by making explicit both parts of a complete explanation. We set out in detail those antecedent happenings which are to be regarded as causes and by stating explicitly the requisite generalizations that justify the relevance of each we leave nothing understood. An explanation in detail of the dying of the fire might be 'Fires die out when their flues are blocked and this flue is blocked'.

'Seeking an explanation' can be understood on the same basis of generalization and relevant particular as 'giving an explanation'. Looking for the explanation of a happening involves both looking for a generalization under which we can include the happening in question, and then with the help of this generalization identifying a cause. A suitable generalization is the prior requirement in seeking an explanation since the identification of any particular happening as the cause of the happening we wish to explain can only be made when the direction in which we are to look for it has thus been established. If we know what sort of happenings are relevant antecedents then we can identify the cause as a happening of this sort. When we try to explain something what we look for first is a generalization under which it can be included. Only then can we identify that special antecedent which is the cause.

This fact has led to some philosophers wishing to say that an event has been explained when it has been subsumed under a general law. Events and laws are the technical analogues of happenings and generalizations. With the latter, as we saw in the study of hyperbolic explanation, there is a definite feeling that the generalization is a prior requirement of explanation. Nevertheless one would not feel inclined, I think, to call an explanation complete or fully satisfactory until one had identified, under the guidance of the generalization, just what happening one would be inclined to accept as the cause. It may be that we shall find in studying science that with the move from particular happenings to kinds of happenings the concept of cause lapses and with it 'explanation in detail'. However, in pre-scientific explanations in which what we seek to explain are particular happenings it is clear that both general and particular causal elements play a part.

If we are looking only for causal explanations then the methods of explanation of particular happenings that we have already discussed can be developed to provide explanations of classes of happenings. In discussing simple explanations I confined the discussion to the explanation of particular happenings for it is in connection with them that our interest in explanations is first excited. However, these methods can readily be adapted to the giving of general explanations, in which we seek to explain not a single particular happening but a class of happenings by stating the kind of happenings which cause them.

I*a*. *General Linear Explanation.* The logical requirements are the same for the general as for the particular linear explanation. An antecedent, preferably the minimum relevant set of conditions for the class of happenings, should be given while the generalizations linking explanatory antecedent and class of happenings are understood. They are implicit in the selection of such and such conditions as the set necessary and sufficient to bring about happenings of the kind to be explained. The requirement that explanation and happening should be of the same logical status is satisfied by our giving as the explanation a class of antecedent happenings. We explain why fires die down by saying 'Their flues get blocked'.

II*a*. *General Hyperbolic Explanation.* The giving of a generalization in explanation of a class of happenings does not differ at first sight from a particular hyperbolic explanation, for a general statement of relevancy can link classes of happenings as readily as particular happenings. But implicit here is a class of events or kind of event which is the class of relevant antecedents of which the particular implicit cause in II was an example.

III*a*. *General Explanation in Detail.* Here we make explicit all that is implicit in I*a* and II*a* by stating both the appropriate generalizations and the class of causal antecedents to a certain kind of happening.

General explanations can be used to explain not only classes of happenings but what I shall call for want of a better expression *generalized types*, things of a certain sort. However, when we ask for explanations of kinds of things there are two ways in which we can understand 'explanation'. For instance such a demand as 'Explain the jet-engine' can be taken in two ways.

(i) Account for the phenomenon of jet-engines. Give an account of what led up to them appearing in the places and in the form that they do.

(ii) Explain and make me understand the jet-engine, that is show me how it works and tell me what sort of thing it is.

When we are asked to explain happenings this question can also be taken in much the same two ways. We may either be asked to account for the happening or to make the questioner understand how it came about. Our three methods of explanation of happenings satisfy both these aspects of explanation at once. We give an antecedent happening to account for the happening to be explained. This will be acceptable as an explanation only if it is the subject of a generalization linking it to the happening to be explained. Now we often say that when we are made aware of the general 'connection' between this kind of cause and this kind of effect we have made an advance in our understanding of the workings of nature. The two elements, cause and generalization, taken together satisfy both our desire to know how to account for the event and our wish to be able to say that we have understood it. When things like jet-engines are in question we can't conflate these demands so easily, for knowing the antecedent causes of jet-engines isn't the same thing as understanding them. In many cases where 'explain' is to be taken as 'make me understand' we turn to analogies.

IV. *Analogical Explanation.* By bringing in an analogy we do not dispense with the formal requirements of explanation but express them in a different way, a way that leads to their being a basis for understanding something. Understanding is usually facilitated by the use of a familiar rather than an unfamiliar mode of expression. In the case of a single happening for example, understanding would obviously be facilitated by the replacement of a relevancy generalization that was expressed in the theoretical terms of an unfamiliar science by one which expressed the same relation concretely in terms with which we were familiar. An explanation of the jet-engine can be given in terms of the analogy with the firing of a bullet from a gun. We are all familiar with the fact that there is a tendency for the gun to go in the opposite direction from the bullet. We have all felt or heard about the kick from a rifle. The analogy can be set up through the equivalences 'bullet is equivalent to the exhaust gases' and 'gun is equivalent to the engine'. Indeed the statement of the facts about recoil vividly expresses the abstractions of the Newtonian Third Law 'Action and reaction are equal and opposite'. If we use the rifle and bullet analogy rather than Newton's Third Law we can make people understand jet-engines much more readily.

V. *Hidden Mechanisms.* A simple causal explanation will only explain one kind of happening. If we want to find explanations that cover many kinds of happenings simple causal explanations will not do. We have to devise a kind of explanation which is simple enough to grasp entire and yet which can be used to give the causes of many different kinds of happenings. The striking of the hours, the movements of the various hands, the ticking noise emitted, and other features of clocks can be explained all at once by describing the mechanism of a clock. Once we understand the mechanism we can state the cause of any one kind of happening on the face of the clock by referring to the relevant part of the mechanism. In this example the hidden mechanism broadens our understanding by explaining many kinds of events.

We can also use hidden mechanisms to broaden our understanding of particular kinds of events. The ticking in the rafters is explained as the outcome of the activities of the death-watch beetle. A particular cause is assigned to a particular effect. But as our understanding of what the death-watch beetle is doing is extended the actual activity which is taken as the cause of the ticking can be narrowed down to various features of its life depending on our particular interest in the matter. It may be that the movement of the mandibles is the cause of the ticking to some well-informed naturalist. For a man called in to renovate the building the mere presence of the beetle will do for the cause of the ticking. An inspector from the Building Standards Office could legitimately settle on poor quality timber as the cause. One simply cannot state, in general, what is *the* cause of the ticking. Hidden mechanism explanations are not themselves causal explanations but rather supply the material for many different causal explanations.

VI. *Explanatory Theory.* As the power and depth of an explanation develops it comes to include the more restricted kinds of explanation we have discussed above as particular cases. Both causal and analogical explanations are included, for by supplying a hidden mechanism of sufficient breadth an explanatory theory can account for many minor causal explanations. From the hidden mechanism we can see just how one sort of happening is relevant to and hence can be the cause of another sort of happening. The mechanism itself is often of such a kind that we can gain an understanding of it most easily through one or more analogies. Daily interests can hardly be said to take us this far. We are commonly satisfied with much less by way of explanation. Most of the things we are puzzled about can be explained either by citing particular antecedent happenings or by the use of simple analogies. Unified explanations, explanatory theories, are the beginnings of science. Any further development of the theory of explanation must be postponed to Chapter Four. In that chapter we shall apply the insight we have gained into simple explanations to the vastly more complicated structures that are used by scientists to account for things that happen.

SYSTEMATIZATION

Any extended discourse contains descriptions of many things and explanations of many things, but these are not thrown together haphazardly. There is always some kind of system in their arrangement. In scientific and mathematical treatises every line and indeed every symbol seems to have its appropriate and ordained place. The statements that occur are arranged in a definite order. Histories have their own kind of system. In poetry one can find systematic arrangements of words and syllables that are in their own way as rigorously designed as the arrangements of symbols in mathematical proofs. A systematic arrangement of words and statements is an ordered arrangement. At the back of any ordering is a principle or principles according to which the ordering is made. In some systems the principles operate directly by giving us rules for arranging statements. In others they operate indirectly by drawing our attention to orderings in nature which, under our principle, we allow to determine our arrangement of statements. Of the many sorts of order possible in discourses we shall discuss only those which have some relevance to the order and system we find in the sciences.

It is sensible and customary to arrange descriptions of events in the same order as the events themselves. When a historical account of something is being given statements are ordered in the same way as the events they describe. This convention eliminates the need for the introduction of the enormous amount of verbal machinery that would be needed to describe the order of events. If the statements describing events are properly ordered we can get along without saying how they are ordered. With the help of clocks and calendars we can give very precise descriptions of the order of events but for practical purposes we do not always need to bring these reference systems into play.

We can derive much practical advantage by ordering descriptions of events in this natural way. When we have produced narratives again and again we sometimes notice that certain successions of events repeat themselves. Such repetitions enable us to make predictions for when we have a sufficient set of examples of a certain repetition we can go on to expect the repetition of the same order again. Our knowledge of such repetitions can be expressed in statements describing the ordering of kinds of events, that is in general statements. When we were discussing the uses of generalizations we saw that we are able to make inferences with them in their conditional form. With the help of generalizations predictions become more than just unvoiced expectations: they become the conclusions of little arguments, guaranteed by our general statements describing the ordering of kinds of events. Novelists and poets conceal the generalizations which determine their ordering of event-statements, for their purpose is to produce descriptions of series of events whose order we cannot immediately infer. Scientists, on the other hand, aiming at the elimination of surprises, do their best to make generalizations about the order of events absolutely clear.

In those arts and sciences which describe events the system is something to a large extent imposed upon the author. There is no free choice about what to say comes first when the events one is describing have a certain order. However, when one is describing things, different and more arbitrary determinations of ordering become important. The spatial ordering of things is often not the best guide to the most useful arrangement of their descriptions. Maps, which show the arrangement of things on the surface of the earth, do not provide us with a stock of useful generalizations except in a very limited way. Comparative geography has nothing like the same precision as chemistry or mineralogy. In what does the difference lie? We were able to use the order in which events occur as the basis for generalization because kinds of events succeed each other fairly regularly. In the natural arrangement of things this is not so. If we pick up every loose object on a country walk the fact that we found this stick after this stone is not of much immediate scientific interest. Classifying things is not assisted to the same extent by their ordering in nature as classifying events is assisted by their natural order. We could easily enough imagine a world in which things of the same kind were always found together but this is not our world. We have the chance to select things in as many ways as we like, a chance denied us in the case of events. A classification of things independent of the order in which we find them in nature is relatively very easy.

The obvious way to classify things is to arrange them in kinds. We put those things together which have certain features in common. We can classify these classes and group these groups too. As we saw in the early part of this chapter the linguistic devices of member and substance generalization are at hand for their description. We can make predictions about things too for we can expect certain combinations of properties to repeat themselves. With the help of the generalizations we use for describing our classifications we can infer from the presence of one or more properties in a thing to the presence of others. If we know that all mammals are warm-blooded we can infer from 'This is a mammal' that 'This is warm-blooded'. Since what properties we select to make our classifications are to some extent arbitrary there are a great many ways in which we can classify things. The classification we come to accept will be that which brings out differences and combinations of properties that are important to us. We don't classify ripe apples with flames, despite their common redness, since it is more important for us to keep the power to burn distinct from nutritive value than it is for us to treat everything of the same colour together.

These methods of ordering descriptive statements depend upon the characteristics of the things and events described and relatively little on the characteristics of the language used to describe them. Descriptive statements can be shown to be logically connected, that is to form the premises and conclusions of arguments, only when there exists a generalization to justify their inference one from another. Any generalization that will allow us to infer new information is one that is based upon some experience of the ordering of events and the coexistence of properties. The arrangement of statements of fact in logical order is dependent in the end on the actual order and arrangement of events and properties. We should not be tempted then to suppose that there is any logic in nature, that one kind of event must follow an event of another kind. The only 'logic of nature' is the collection of generalizations and principles we call the Laws of Nature, and they depend upon the nature of the world.

There is another kind of ordering of statements that is very important in the sciences though it does not allow us to infer new pieces of information. This is ordering according to the general rules of logic, and, especially in science, the rules of mathematics. For example if metals are taken to be divided into two groups, paramagnetic and diamagnetic, then we can infer from 'All metals are conductors' both 'Paramagnetic metals are conductors' and 'Diamagnetic metals are conductors' without even knowing what the words 'paramagnetic' and 'diamagnetic' mean. The two statements, 'Metals are either paramagnetic or diamagnetic' and 'All metals are conductors', already contain all the information which is contained in 'paramagnetic metals are conductors' and 'diamagnetic metals are conductors' and more than is contained in each singly. In one way this is an empty move yet it may sometimes be of the greatest importance to bring out these two separate facts about kinds of metals clearly and distinctly. Furthermore the generalization upon which the inference of the two concluded statements is based is not one to be learned by watching and collecting information about events or things. It is of a different nature. 'If "All X are Y" and "X is A plus B" then "A is Y" and "B is Y"' is a conditional statement which if accepted will guarantee our particular inference above and an indefinite number like it. The generalization which is equivalent to this conditional will give us the clue to the kind of inference this is for it must be expressed in terms of the form of the sentences used, not their content. The generalization is 'When any statements made with sentences of the form "All X are Y" and "X is A plus B" are true then statements made with sentences of the form "*A* is *Y*" and "*B* is *Y*" are also true'. This is one of the rules of logic. There are many different kinds of rules in logic but they all have this in common; that they are concerned with stating and establishing connections between the forms of the sentences used to make particular statements, so that through infinite changes of form truth is preserved. The forms with which we start are called *axioms* and the forms we deduce from them *theorems*.

In the example we have been discussing the move we have made with the use of our formal rule has been from the more to the less general, but we gained no addition to our knowledge of the world by making this move. We simply changed the form of our knowledge. For certain purposes it may be more convenient to have our knowledge in a different form. This device of using axioms to contain all the more specific forms of knowledge is of enormous value as we can, by picking our original forms skillfully, represent whole fields of learning in a few basic principles. By applying the rules of logic appropriate to the sort of axioms we have any specific item of information can be deduced. In this way we reduce our knowledge to deductive systems. This method of forming systems is quite different from the two we examined above, for those systems depended for their structure upon the actual order of events and our classification on the real properties of things. Deductive systems do not depend for their structure upon either of these features of the world but rather upon the form into which our knowledge of the world is cast. The first great deductive system was the geometry of Euclid, but its influence was not felt in the other sciences for a very long time. The proper form into which to cast our information about events and the properties of things (as distinct from their arrangement) was very difficult to discover.

Our study of the methods by which scientists describe things and events, explain happenings and systematize their knowledge will lead us to see that it has been one of the fundamental aims of science to reach deductive systems of knowledge. It is hoped that systems will be found which are powerful enough to include and explain all the more minor systems of classification which depend directly upon the character of the world.

REFERENCES

1. P. F. Strawson, *Introduction to Logical Theory*, pp. 3, 4.
2. G. Frege, *The Philosophic Writings of Gottlob Frege*, trans. P. Geach and M. Black, p. 21 ff.
3. C. S. Peirce, *The Simplest Mathematics*, p. 423, Sect. 537.
4. P. F. Strawson, 'Particular and General', *Proc. Aristot. Soc.*, p. 238, (1953/54).
5. J. Watling, 'The Problem of Contrary-to-Fact Conditionals', *Analysis*, **17**, No. 4, p. 73 (1957), this article puts the main features of what has been thought to be the problem. See also N. Goodman, *Fact, Fiction and Forecast*, and the series of articles in *Analysis*: S. Hampshire, 'Subjunctive Conditionals': **9**, No. 1 (1948); D. Pears, 'Hypotheticals', **10**, No. 3 (1950); W. Kneale, 'Natural Laws and the Contrary to Fact Conditional', **10**, No. 6 (1950).
6. R. M. Chisholm, 'The Contrary-to-Fact Conditional', *Mind*, October 1946; F. L. Will, 'The Contrary-to-Fact Conditional', *Mind*, July 1947.
7. S. Kripke. 'Semantic considerations on modal logic', *Acta Philosophica Fennica*, **16** (1963), 83–94; and D. Lewis, *Counterfactuals*, Blackwell, Oxford, 1973.
8. B. Mayo, 'Conditional Statements', *The Philosophical Review*, **LXVI**, No. 3 (July 1957).
9. J. L. Mackie, *The Cement of the Universe*, Oxford University Press, 1980.

2 Interlude: The Aims and Methods of Science

The needs which have produced science are not something new, that suddenly inspired exceptional people to undertake a new kind of activity. They are indeed very ordinary, springing from the general insecurity of human life. Security is enhanced when from an analysis of the situation in which we find ourselves we can make intelligent anticipations of the likely future. I do not want to pursue speculation into the anthropological background of science into any great detail but it is helpful in understanding the complex business of modern scientific procedures if we remind ourselves from time to time of these anthropological truisms.

Knowing how to manipulate our environment to our own advantage gives us the power upon which our general security can be based. This ability can be achieved in a limited way by simply generalizing the information which we already have about the world so that predictions of the future course of events can be made by inference from what is the case now. However, much more can be achieved if we can say that we understand the workings of the world, if we know why and not merely that such and such an event will come to pass. Simple generalization will not achieve this further aim, and it is as a response to the need for explanation that the theoretical structures which are so characteristic of modern science have been developed. These facts have not always been clearly seen. The history of the philosophy of science could be written as an antiphony between philosophic views appropriate to the collection of superficially related generalizations to those appropriate to the attempts to build explanations of such generality and power that at least within certain restricted classes of phenomena nothing is unexplained and everything can be foretold.

Bacon's handbook on method, the *Novum Organon* (1620), presents both a logical analysis and a philosophic justification of the procedures he recommends. This analysis and justification serves to strengthen the attractiveness of the method. Bacon hoped to show that his method, by contrast with current views of the time on methodology, was the only one with which true scientific progress could be made. For Bacon this meant a method by which the natures of things could be discovered without falling into any of the errors which prevent the acquisition of knowledge. It was characteristic of early work on the philosophy of science that great attention should have been paid to the justification of the means of augmenting human power. Bacon says[1] 'Upon a given body to generate and superinduce a new Nature or new Natures is the work and aim of Human Power'; and 'the true and legitimate goal of the Sciences is none but this: that human life be enriched with new discoveries and wealth'[2]. These desirable consequences are dependent upon our having the requisite knowledge—'To discover the Form of a given Nature, or its true Difference, or its causal Nature, or fount of its emanation . . . this is the work and aim of Human Knowledge'[3]. The way to this knowledge is the consultation of nature—'In experience of every kind, first the discovery of causes and true Axioms is to be made; and light-bringing not fruit-bringing experiments to be sought for'[4]. Once we have this knowledge the practical applications can look after themselves for as Bacon says[5] '. . . Axioms rightly discovered and established supply practical uses, not scantily but in crowds, and draw after themselves bands and troops of effects'. But the acquisition of knowledge must be systematic and based upon the extension by means of hypotheses of knowledge already acquired[6]. 'The true order of experience first lights a torch, then points out the way by means of the light, by beginning with well-ordered and digested experience, which is nohow misplaced or vague, and by drawing out thence axioms, and from axioms when established new experiments. . . .'

Bacon was suspicious of deductive systems justified only by the light of reason. His method of trial and error was aimed at providing well-grounded theories of the essences or inner natures of things, by which Bacon understood the 'latent or hidden configurations' of their constituents, which might, for example be particles. Modern chemistry is perhaps the nearest any of the actual sciences have come to fulfilling Bacon's methodological dream.

At about the same time the development of the medieval theory of proportions into a powerful method for the mathematical description of motions and structures led to the use of a geometrical rhetoric to write up scientific work. Newton sets out both his *Principia* and his *Opticks* as axioms, theorems, corollaries, etc. in just the style of Euclid's *Elements*. But the *Opticks* is certainly not a deductive system, and the relation of its sequential steps is not achieved by anything like a geometrical proof. The *Principia* looks more like a deductive system but it is at least possible to interpret Newton's Laws as rules for the use of certain mechanical concepts which determine how these concepts are to be applied, concepts such as acceleration, force and so on, in particular cases. On this interpretation the *Principia* is a kind of illustrated dictionary or encyclopedia.

A growing interest in the relation of experience to knowledge led to Locke's influential distinction between ideas and qualities, that which was in the mind and that which was in the world. The science of mechanics was possible, according to Locke, because some ideas resembled the real qualities of things (the primary qualities of bulk, figure, texture and motion) though many did not (ideas of secondary qualities such as feelings of warmth and sensations of colour). Furthermore he though Newton's work had demonstrated the rightness of extending the scope of primary qualities to include the fine structures of things (Bacon's latent configurations), structures which were beyond the powers of the human senses to experience.

This optimistic 'mechanization of the world picture' was challenged by the scepticism of David Hume, whose views have provided a perennial source of arguments against the pretensions of science to reveal the hidden workings of nature.

Hume had argued that all factual propositions were reduceable in the last analysis to statements about simple impressions, and that the most that could be said about their connection was that different kinds of impressions succeeded one another in some restrictedly systematic ways. The apparent necessity of these connections, is imposed by us under the influence of the force of habit. So there is no logical force in predictions of the future. Indeed Hume went so far as to say that only those propositions which are developed systematically without application to the world can be taken to be certain. These, in practice, reduce to such statements as the rules of logic and the axioms and theorems of mathematics.

It was to counter this scepticism that Kant developed his theory of the 'synthetic *a priori*' propositions. There were certain fundamental propositions upon which Newtonian science was thought to depend; for example that a given event must have a cause. Hume had argued that in this example the word 'must' is to be understood as representing no more than a certain habit of mind formed by experience. Kant, being concerned to show that the connection was a stronger one, thought to argue that the proposition in question, while applying to the world and so descriptive of nature, was imposed by us upon our experience, and so in a way conventional. Thus he was able to say that the general laws of nature are both synthetic and *a priori*. Of course the conventionality is not arbitrariness, and the *Critique of Pure Reason* is largely concerned with showing that the general laws of nature and some other very general propositions are required, logically required, as the conditions of our having any experience at all; at the very least experience discriminable in the way we know it. This contrast between Humean scepticism and Kantian 'conventionalism' has survived into modern controversies and is to be found in the differences of the schools of philosophic opinions of this century. This fact alone should lead us to suspect that each has reached the position they have through an overemphatic insistence on one aspect of the scientific process.

A revival of the Humean view of science as a collection of rules of thumb grounded in habit began in continental Europe at the end of the nineteenth century, particularly in the work of Ernst Mach and his followers. Mach said (1894), 'The communication of scientific knowledge always involves description, that is a mimetic reproduction of facts in thought the object of which is to replace and save the trouble of new experience. This is really all that natural laws are. Knowing the value of the acceleration of gravity and Galileo's laws of descent, we possess simple and compendious directions for reproducing in thought all possible motions of falling bodies. A formula of this kind is a complete substitute for a full table of motions of descent, because by means of the formula the data of such a table can be easily constructed at a moments notice without the least burdening of the memory. . . . More than this comprehensive and condensed report is not contained in a natural law of this sort'[7]. Mach goes on to say that when this process of generalization is complete a certain province of facts has been *explained*.

This view of science has been called 'positivism', and in the extreme form given it by Mach there is no place left in science for the understanding of nature beyond its representation. Clearly many important scientific advances have occurred when beyond the immediately perceptible facts which can be related as Mach recommends, some explanatory hypothesis postulates an entity or realm of entities waiting to be discovered; and for their properties to be investigated. In the fully developed positivism of those philosophers who formed the Vienna Circle, intervening theory was reduced to a kind of inference-machine and its postulated elements to logical constructions out of the immediately given. I do not propose to investigate the group of theories that developed round this central idea in detail, but it is important to see that the positivism they recommended takes the meaning of the expressions and the statements appearing in science to be wholly determined by the observational material out of which they are 'constructed'. As it stands it is an exaggeration and tended to force those who held it into untenable positions, but it has this merit, that it brings before our attention the overriding importance of facts in the testing of theories which a Kantian 'conventionalism' can obscure. It was by insisting upon the role of 'facts' in tests that Popper turned the positivism of the Vienna Circle upside down, with his doctrine of conjectures and refutations.

The recent history of the philosophy of science can be partially understood as a series of controversies deriving from the two elements in the thinking of the philosophers of the Vienna Circle and those influenced by them. On the one hand the use of the concepts and relations of formal logic as the ultimate analytical tools for understanding scientific discourse has been challenged by those who are concerned to see the rationality of science displayed in its actual practice (for example one can look on the works of Feyerabend and of Laudan in that way) and by those who believe that that rationality is to be found at least partly, in the content rather than the form of scientific discourses. (For example both Bunge and Toulmin have taken this tack.) So there are, so to speak, two opposition parties to the dispute, who are themselves in conflict. The second axis of polar oppositions turns on the anti-realism of the logical positivists, that is on their positivistic stance. Again there has been a heterogeneous collection of opponents, including Popper who tried to combine faith in realism with a commitment to logicism, and Newton-

Smith who tends to look to the actual practice, the history of science, for support for the realist position[8].

The analysis of scientific thought and practice in this book is in the tradition of those who believe that some scientific theories do successfully and legitimately go beyond experience to reveal the unobservable, by acts of thought; and of those who see the 'logic' of scientific thinking as a more enlarged version of rules for rational thought than are encompassed in formal logic alone. In the chapter that follows I hope to give an account of science that tends to neither extreme, for I believe that among the sciences we can find examples of theory construction which form a kind of spectrum. There are those which collate information by Baconian induction; those that collate their information more broadly by the introduction of a general explanatory model, and there are theories from which all that is pictorial is purged leaving only the representation of the immediately observable and the formal inter-relations of symbols. I do not know whether in this matter ontogeny recapitulates phylogeny, it may indeed be so. The reason why we are inclined to call a science that depends for its explanatory power on nothing more than a schematized description of the facts in the field in which it works an undeveloped or rudimentary science may be that progress in physics and chemistry has been towards a more and more advanced formalism. What is certainly true is that among contemporary sciences a spectrum can be found in which at one end they are purely descriptive and at the other dully formalized.

REFERENCES

1. F. Bacon, *Novum Organon*, trans. G. W. Kitchin, p. 113.
2. *Loc. cit.*, p. 57.
3. *Loc. cit.*, p. 113.
4. *Loc. cit.*, p. 45.
5. *Ibid.*
6. *Loc. cit.*, p. 60.
7. E. Mach, *Scientific Lectures*, trans. T. J. MacCormack, pp. 192–3.
8. For references to the 'rationality of science' debate, cf. p. 175.

3 Scientific Description

In Chapter One a general classification of descriptions was undertaken based upon whether in the description a property was assigned to an individuated object or to a substance. The sciences are concerned with something still more general, which I shall call, borrowing Margenau's terminology[1], *systems*. As I shall use the word any extended object which is so organized as to maintain its integrity for some time is a system—the universe, an atom, a bird, an ecological region, the fish population of the North Sea, a radio receiver, are all systems. I shall say that a basic aim of science is to give adequate descriptions of nature as a hierarchical open set of systems. Adequacy of a description will depend on several criteria, correctness or accuracy being only one, though there are a collection of general standards to which any particular judgement of adequacy must be referred.

A system has properties and an internal structure; that is it has a certain character and it is built up in a definite way out of sub-systems. Scientific description aims at (i) specifying the structure of the system, by describing the sub-systems and their arrangement and connection. These sub-systems may, of course, become the subject of structural description themselves. (ii) Science aims at specifying the properties of systems, using the fewest possible predicates. (An aim of scientific explanation is the expression of properties of systems in terms of the structure of the system and the properties of the sub-systems which make it up.) A property of a system which is selected by a descriptive predicate in the minimum group I shall call, following the accepted terminology, an *observable*. The complex of observables, that may at any time be the properties of a system I shall call the *state* of the system at that time. With the introduction of the notion 'state of a system' a new dimension of description is opened, since a given system may be in a certain state at one time and in a different state at another. We can aim at describing the change of state that has occurred between the two times at which the state of the system was described. Further it is here that the basic aim of scientific explanation enters, to give an account of the conditions, the antecedent states of the system and the external influences upon it, that lead from the one state to the other. The two kinds of descriptions that can be given of a system, structure and change of state, are clearly distinguished in some sciences; for example in the description of biological systems, plants and animals, structural description is given in anatomy, and change of state description in physiology.

There are a great many different kinds of properties that systems exhibit and a selection is made among them for those which are to be chosen as observables, in which the states of the systems will be expressed. In general in giving a scientific description a property is selected as an observable if it can be quantified and measured, that is expressed in terms of a cardinal number and a unit, the unit being such that operations for its determination can be described. These are not the only determining factors in the selection of observables since the demands of explanation are also determinants as we shall see in the next chapter, but as far as description is concerned a quantifiable and measurable property will be selected as an observable rather than a property not so expressible. For example the qualitative property expressed in the predicate '. . . is hot' is replaced for the purposes of scientific description by the observable expressed in the predicate '. . . has a temperature of $x°$ C'. The reasons for this replacement are complex and will become apparent as the discussion proceeds.

Static anatomical descriptions of systems require no special linguistic machinery apart from that already deployed in ordinary language, but the dynamic description required to express change of state does not have so direct a logic. A dynamic description could be built up out of static descriptions in the following way: the system is described at some time t_1 and again at some time t_2. A dynamic description is built up by saying that the system is changed from a state S_1 which was characteristic of it at t_1 to a state of s_2 which was characteristic of it at t_2. This method of dynamic description is built up by saying that the system cumbersome if a very detailed history is contemplated or if an attempt is made to express continuous change of state. A way out of the difficulty can be found by the use of a very fruitful and important analogy.

If we describe the static state of a system by means of quantified observables we express the results of our measurements as quantities, that is as products of cardinal numbers and unit expressions, the former expressing the number of operations of measurement defined by the latter, for example '11 × 1 inch'. Change of state occurs when one or more observables change their values, that is require a different number of operations for their measurement. There is a mathematical device for the representation of a number without expressly saying which, the algebraic number-variable. A changing observable can be represented by an algebraic number-variable with a unit expression. The velocity of an accelerating body, different at different times, can be represented in general by 'v ft. per sec.'. If description is in terms of the same set of observables, particular systems and kinds of systems will differ in the way the observables change with respect to each other. To this feature of changing states there is also an algebraic analogue, the function of variables. For example the variables x and y can vary in infinitely many ways with respect to each other, one such way being represented by the function

$$y = 2x$$

An analogue in a physical system can easily be imagined. For example the blowing up of a cylindrical balloon which is made in such a way that the length is always twice the diameter provides an analogy. The change of state of the system 'balloon' can be represented by the above function if y is taken as the length-variable and x as the diameter-variable. A change of state of a specific kind will be in general be called a *process*, as for example we can say that the function above represents the process of expansion of such and such a balloon.

This method of description can be generalized. Let $x, y, \ldots$ be algebraic variables representing observables, and $a, b \ldots$ be numbers representing the results of measurements on the observables represented by $x, y, \ldots$ respectively. Then if $F(\ldots)$ is an algebraic function representing the process f which the system is undergoing,

$$F(a, b, \ldots)$$

represents a particular state of the system, a static description true at some particular time. The same function, but with variables instead of definite numbers

$$F(x, y, \ldots)$$

represents the change of state of the system, the process f, a dynamic description of the history of the system.

The analogy is usually pressed a stage further, and turned back upon itself, by our talking of the observables of the system showing a certain functional dependence. Correlation of observables with algebraic variables is not always possible, for example in cases where no method of quantifying a property is known. Sometimes it may not be desirable where quantization of one variable-observable may give a false impression of the precision of description of a state or process in a system, as in certain descriptions in the social sciences. There is indeed in the relation and dependence of variable factors a range of exactness which varies from that of commonplace sociological generalizations like 'The more the merrier' through more or less precisely quantized variables to the exact functions of physics, of which the algebraic functions are typical. I do not mean here to claim that observables are exactly determinable in physics, but that the form of descriptions is exact and not vague, whereas in the social sciences both the data and the form of expression may contain unavoidable inexactnesses. The natural method of describing the history of a system, rejected above for its clumsiness, can be seen to be included in the functional method of description, for by the substitution of values of variables appropriate to a certain state, descriptions of the state of the system can be reproduced for any point in its history. The method of representation, which might be thought of as the ideal of scientific description deserves a closer analysis.

The functional generalization that has been described in general above is tied to measurement, the measurement of observables. The formation of the description must be linked with the carrying out of experiments with apparatus designed to record in numerical terms the observables of a system. Suppose that we distinguish a use of a particular apparatus under particular conditions as an experimental-situation. The further use of a particular kind of apparatus under differing conditions can be represented by a set of experimental-situations $s_1 \ldots s_n$. Let k experiments be carried out in a given experimental-situation, so that the class of experiments of this kind will be $e_1 s_1 \ldots e_k s_1, e_1 s_2 \ldots$

$e_k s_2, \ldots e_1 s_n \ldots e_k s_n$. A complete generalization independent both of a particular experiment and of a particular experimental-situation can be obtained in two steps.

1. In each experimental-situation we can generalize the results obtained from the k experiments into a functional expression representing the change of state of the system under investigation. As an example it can be supposed that a measurement of current and voltage in a conductor is being made, the correct functional expression being found from the numerical results obtained. It is found that the connection is a simple one, voltage varies with current. The numerical results for each quantity can be generalized into a variable, say V for voltage and I for current. We can enunciate the functional generalization

$$V \text{ varies as } I$$

However, we can also see how they vary, what the exact connection is between them. And for a given experimental-situation a definite value for a numerical parameter can be calculated. It may be that in a particular circuit the conductor is such that the numerical value of the current is always twice that of the voltage.

2. In another experimental-situation s_2 we find that current is always three times the voltage, and so on, each set of experiments $e_1 s_1 \ldots e_k s_1, \ldots, e_1 s_n \ldots e_k s_n$, giving a functional law of the same form but differing in numerical constant. At this point a second stage of generalization can be undertaken, but now two moves are open to us.

(*a*) The numerical constant, differing with each experimental-situation can be generalized into an algebraical variable, an experimental parameter, and we say that it represents a special property of the experimental-situation. For example the experimental parameter obtained from the above experiments on the variation of current and voltage in a conductor is called the resistance of the conductor and is adopted as an additional property of the experimental-situation. The result of making this move is the function

$$V = IR$$

where R is the parameter for the experimental-situation.

(*b*) The variation of the parameter with experiment can be the subject of further experiments aimed at linking its variation with the variation in some other, hitherto secondary property of the apparatus or system; or with the variations in the conditions that had been absorbed into the background of the experimental-situation as irrelevant to the variation in the original observables chosen for experimental study. For example we may decide that the length and properties of the conductors in the apparatus are relevant to the differences in parameter, or that the temperature conditions are worth investigating. What we do in fact pick on for further study is closely connected with the explanatory theory that exists for this field of study. I shall return to this topic in the sixth chapter.

Each of the possible moves, (*a*) or (*b*), is in fact fruitful of new generalizations. Treating the parameter as itself a physical quantity the laws of combination of resistances for example can be investigated, while the kind of investigations described in (*b*) above led to the discovery of connections between resistance, cross-section, material, length of conductor, temperature and so on. The first step in generalization when variables are introduced so that the expression of the process investigated is independent of a particular experiment, but not of a particular experimental-situation I shall call *immediate* generalization. The second step in which by either method the generalization is freed from numerical dependence upon a particular experimental-situation I shall call *parametric* generalization. Measuring techniques and the mathematical analogy combine to determine the form of the kind of generalization that has become the ideal in scientific investigations. It will become clear below that the advantages that accrue from this type of generalization are not just precision of representation and economy of description.

It is possible to some extent to represent those cases mentioned in the general discussion above in which no precise connection between the variables representing observables can be immediately determined. In fact exact generalizations represented by continuous algebraic functions like $V = IR$ can be thought of as limiting cases of much less precise relationships. Consider again the method of generation of the relationship $V = IR$. In a given experimental-situation the results of experiments could be tabulated. For example

Voltage: $v_1, v_2, v_3, \ldots$
Current: $i_1, i_2, i_3, \ldots$

The function $V = IR$ is built up in two steps, one is the generalizing of the numerical results $v_1, v_2, \ldots$ into the variable V, and $i_1, i_2, \ldots$ into the variable I; the other the determination of the exact relationship between them, that is the determination of the algebraic function which has numerical values varying in an analogous way with $v_1, v_2, \ldots$ and $i_1, i_2. \ldots$ In this example it is found that in a particular experimental situation the quotients $v_1/i_1, v_2/i_2, \ldots$ are the same, and this can be represented via the first step as the quotient V/I. Whatever the list of results tabulated represents, the first step in generalization is possible, that is the representation of each set of results in an algebraic variable, but in many cases the second step is not. For example it might be found that all the immediately obvious arithmetical operations with the numerical values of the observables failed to suggest a relationship. This would not mean that the observables were necessarily entirely independent in their variations. Dependence that falls short of functional representation can be expressed by an index of correlation, the details of the calculation of which are a special study in statistics and need not detain us here. The statistical measure of correlation does not, of course, allow an exact functional connection to be established, but it points to where one might be found if the correlation between variables is good. If the correlation is poor it shows that further investigation is very unlikely to reveal a functional connection. If the correlation is good but not perfect we must postulate the interaction of other variables, and by taking into account previously unconsidered factors in the situation—observables in the system—exact generalizations may well be found.

Parallel to the algebraic analogy goes a graphical analogy with the variation of observables in the process on systems. Indeed in some respects graphical generalization is more powerful than functional generalization for by means of graphs we can represent correlations too complex to be expressed analytically. The same steps, from data to immediate generalization and from immediate generalization to parametric generalization can be traced in the construction of graphs. To invent a simple illustration; suppose that the following table of results has been obtained from an experimental-situation *e-s A*.

V	1	2	3	4	5	6	7	8	9	10
I	2	4	6	8	10	12	14	16	18	20

And the following table from an experimental-situation *e-s B*.

V	1	2	3	4	5	6	7	8	9	10
I	3	6	9	12	15	18	21	24	27	30

Graphical representation of each variable along axes mutually at right angles will allow the representation of the mutual variation of the observables.

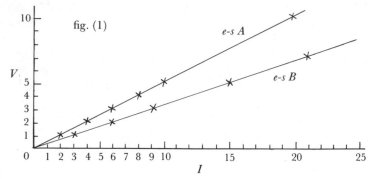

Graphs *A* and *B* differ in slope, so that we may select slope as the experimental-situation parameter. Generalizing this we dispense with numerical expressions altogether and thus with definiteness of slope, as in fig. ii,

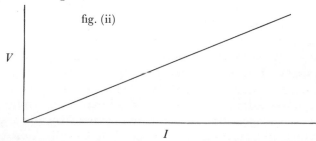

contenting ourselves with representing the form of the curve, in this case a straight line in the first quadrant. No particular slope is specified and so the graph represents the type of process of the observables and is independent of particular experimental-situations. It is customary to distinguish the particular expressions of fig. i above as the results of graph-drawing and the general in fig. ii as curve-tracing. In the presentation of the results of real investigations graphical methods are extremely important, descriptions of systems sometimes being confined to them. Their systematization, however, is difficult and as we shall see below it is important that the corresponding algebraic expressions should, if possible, be discovered.

The methods of description outlined above are idealizations of the actual procedure of the sciences, and furthermore they are characteristic of a late stage in the investigation of a field of enquiry, since they require that the relevant observables should have been identified and that they are either themselves measurable or can be expressed in terms of observables that are. The analysis and classification of systems and sub-systems, that is their detailed description is an essential preliminary to any quantitative determinations of observables. A preliminary description in terms of qualitative, or at least unquantified predicates is required both to delimit the field of investigation to the relevant observables and to identify the grosser connections between the properties of systems. The preliminary pre-mathematical stages of description can be found in procedures which are not now characteristic of physics but can be found in other sciences.

Cumulative Generalizations. Generalizations made in certain biological experiments are typical of methods of description and generalization different from the method of functional dependence; for in many biological studies it is not a dynamic system that is under investigation. The first case I want to consider is typified by the following simplified example. Suppose a study were being made of the effect of a certain drug (X_2Y) on animal health. The method usually employed in this kind of investigation is to obtain a large number of specimens of the animal to be investigated, say from the mythical species *animalis innominata* and to divide these into two groups, one group to be selected for medication and the other to be retained as a control. The initial state of all individuals in both groups is arranged to be as nearly the same as possible, and after the experiment is begun the onset of symptoms is looked for in both groups. The course of the imaginary experiment might run as follows:

Medicated Group: Of 30 animals 27 had developed hypernormal temperatures by the third day.

Control Group: No abnormal temperatures were observed. From the results of this simplified investigation a general statement can be made:

'Hypernormal temperatures follow the administration of $X_2 Y$ 27 times out of 30 specimens of *animalis innominata*.'

Several logical points emerge from studying this procedure. It is to be noticed that in this kind of experiment there are what I shall call *countables*, discrete individuals, which serve to provide the quantitative basis of the experiment by their being simply counted. In contrast to the functional generalization which describes the history of a single system, the generalization above is extended, covering many individual systems, single animals, the purpose of the extension being to provide countables. The generalization which finally emerges is cumulative in that it sums up the results of many sub-experiments on many individuals. It is cumulative generalizations of this sort that provide the clearest cases for the application of probability theory.

A preliminary point[2] about the application of probability-measures to generalization can be made here with the help of the above example. The evidence for a correlation between administration of $X_2 Y$ and the increase in body temperature is not absolutely conclusive, for three of the medicated specimens did not react to the drug. In stating the results of the experiment in a generalization this fact can be included in two ways.

(i) The generalization can be unrestricted in form, with an attached rider stating that the evidence for it is not conclusive; that is the results can be stated in an unrestricted generalization with restricted support.

(ii) The generalization can be restricted in form, with a rider attached stating that the evidence for it is conclusive. These cases are conflated in the crude expression of probability.

'Probably the administration of $X_2 Y$ to animals of the species *Animalis innominata* is accompanied by a hyper-temperature', can be read in either way. Clearly there can be no doubts about confirmation in case (ii) for no more is stated in the generalization than is given in the data, while in (i) the generalization being unrestricted, is for that reason uncertain on the basis of the available data. A similar point can be made even when every experimental result is positive, provided that as commonly occurs, there are cases which lie outside the range of those actually examined. The connection between generalization and evidence in cases of this sort is of great methodological importance since one important means of scientific inference is from evidence to a hitherto unknown generalization. This is very rarely a simple inference of the type of case (ii) above, and its detailed study will be undertaken in Part Two. At this point it is important only to see that cumulative generalizations and their evidence, the information which they 'accumulate' are not of the same logical status and that when an unrestricted generalization is attempted some disclaimer of complete support must be made. The standard disclaimer is made through the notion of probability.

In the experiment above the determination of a result was simplified to a simple query, which required a positive or negative answer depending upon the possession or non-possession of a property. In the example, this was the possession of a hypernormal temperature by each individual specimen. This represents a limiting case, for in practice the property taken as an observable is quantifiable and so a *measurable*. Most experiments are not so simple as that above but are characterized by quantization through both countables and measurables. For example the simple query above could have been supplemented by a determination in the standard quantitative way of the degree of rise in temperature, and a further dimension of facts ascertained. It might be the case that while every animal in the medicated group exhibited some degree of hypernormal temperature, there were different maxima reached. From such facts as these investigators are led to further experiments to try to correlate such fluctuations with other factors in the situation, and so by degrees to a functional study of an individual system.

Non-cumulative Generalization. Generalization was accomplished in the above example by the device of investigating all the members of a range of individuals. Another device which is commonly employed is to pick on a certain specimen as the *type* of a range and investigate just this specimen, generalizing by stating the results of the investigation as predicates of a species or in other cases of a substance. This method of investigation and description I will call *non-cumulative* generalization.

Typical cases of non-cumulative generalizations occur in those sciences that are concerned with substances; and in particular in chemistry, both qualitative and physical. Such a generalization as 'Gold has a density of 19.6 gm. per c.c.' or 'Gold is soluble in *aqua regia*' may be used for one of two different purposes. It may be used to state the defining properties of the substance, properties that could be used by an assayer to distinguish gold from other yellowish metals. It may also be used to state the results of new investigations of the substance, gold, investigations which presuppose that the identification based upon the defining properties has been carried out. These investigations may quite properly be carried out on single pieces of gold provided that certain safeguards have been observed. The necessity for safeguards follows from the methodological doubts expressed by 'How can the investigator be sure that the specimen he is using to determine the properties of the substance is a genuine representative?' and 'What sort of statement is that which sets out the defining properties of a substance, how is it related to the facts?' These questions are connected, since what guarantees the specimen to be a genuine representative is the presence of the defining properties together with a standard process of specimen selection.

In pre-scientific metallurgy the rough set of defining properties which were necessary to recognize gold could be unspecified and unspecific. So long as no serious difficulties of identification and misidentification arose a vaguely defined set of properties would be quite satisfactory, provided that it was specific enough for current practical purposes. The story of Archimedes and the King's crown illustrates the development of specific criteria for the method he used to test whether the gold was pure was to use specific gravity as a uniquely determining characteristic. It is important to see that once the selection of a property or set of properties from the vaguely defined group which are habitually used for identification of a substance is made, then the attribution of this property or set of properties to the substance cannot be false, and the identification of a specimen as a genuine specimen by means of these properties cannot be a misidentification. The statement expressing the attribution of these properties to the substance is clearly an analytic statement and the use of these properties allows for the identification of an instance as genuinely representative of the substance. Experiments with such a specimen do not require an accumulation of instances from the investigation of other pieces of gold before a generalization can be announced; the statement of the results of experiments on such genuine specimens can themselves serve as non-cumulative generalizations.

There are two reservations which must be kept in mind. It is not, in general, possible to say once and for all what the limits to the set of defining properties may be. Suppose a counterfeiter produced an alloy which was yellow, malleable and ductile, of density 19.6 gm. per c.c., insoluble in all but *aqua regia*. Suspicion of the substance could lead to an extension of the set of defining properties, if they were the currently accepted set. We could perhaps include 'having a spectrum of such and such a character'. Moves by the counterfeiter to simulate each additional property could be countered by an addition to the range. However, should there be no further property which can be added to the list that the alloy does not simulate then we must grant that the counterfeiter has stumbled on the philosopher's stone and produced gold. It is seldom in practice that the matter rests here; for example it is said that such is the 'genuineness' of artificially produced rubies and amethysts that only if a ruby has a flaw is it highly prized.

The second reservation is concerned rather with the rest of the substances than with the specimen by which it is being investigated. It is this—how can we guarantee the uniformity of the substance, on what do we base the belief that a substance, wherever an instance is taken, will follow the defined instance used in the determination of the new property exactly? In practice this requirement is met by the process of purification. It is assumed in chemistry that differences in the properties of specimens taken from different parts of the same substance can be accounted for either by differences in their physical properties, *e.g.* whether in crystalline or amorphous form; or by their having undetected impurities. Uniformity of behaviour can be ensured according to chemical practice by ensuring that the processes involved in the extraction and preparation of the substance are uniformly carried out. These processes can be summed up as purification; and there are standards laid down as to what constitutes a 'pure' specimen of any given substance. These standards vary in different contexts, the standard of purity required of domestic common salt differing from that required of sodium chloride, prepared for use in a laboratory. The standards of purification differ from the defining properties in that these standards can be given in operational terms, specifying the series of processes that the substance must undergo from the finding of the ore to the bottling of the specimen. Standards of purity apply to operations of purification while the defining properties apply to the substance itself.

The application of analogues of the non-cumulative procedures to range-of-individuals generalizations occurs in biological descriptions where a particular plant or animal is investigated as a type of the species. We investigate *Lepus caniculus*, the rabbit, in the persons of particular individual rabbits, and we do not need to repeat the investigation which gives us a certain piece of information in a multiplicity of cases before we announce this information as applying to the species, provided that we are satisfied that the specimen actually chosen is typical. The guarantees of this are analogues of the guarantees that are used in chemistry. The statements of the characteristic features of a species cannot be false in the same way as the statement of the defining properties of a substance cannot be false; and though the skilful recognition of individual peculiarities that are to be discounted in a specimen is a much less precisely expressible thing than the operations of purification of substances there is a close logical parallel.*

Of course this convenient device of non-cumulative generalization is not entirely independent of the accumulation of instances. Individual differences in the members of an analytically defined species that are discovered by the accumulation of information to be systematically repeated do lead to divisions of species into sub-species. Serious problems in taxonomy occur in deciding when an individual difference is sufficiently important or sufficiently widespread in the members of the species to be regarded as defining a new species. When an affirmative decision is made in such a case the ascription of the property to the new species becomes an analytic statement, serving to express the decision that this property constitutes the major differentium of the species. Minor individual differences are 'leached out' of a specimen by the common device of schematization, in which only those features which are relevant to the species are retained and nothing that is relevant to the recognition of this individual from others of the same species remains. This is easier to recommend than to do, and both considerable experience of biological investigation and a wide cumulative knowledge of the species is necessary before a useful and exact schematization can be made. This process of schematization is one of the beginnings of theory and I shall

*There is this difference, however; biologists, particularly botanists, use actual specimens rather than written definitions.

return to it below. At this point it is relevant only as a way of presenting non-cumulative generalizations.

But prior to this intrusion of accumulated differences the cumulative generalization plays an important part. Non-cumulative investigations depend upon our selecting a particular individual as a specimen, representing the type of all the instances of the substance or the species. To make this selection we need already to know those properties which typify the species, and these can be arrived at in different ways. They might be laid down by fiat as if we were designing a world and inventing a language to go with it, in which case it would be merely fortuitous that a species could be found answering to the *a priori* description. On the other hand by the examination of many animals, roughly sorted into species to begin with, the really typical form and its appropriate description gradually emerges. In practice the methodological procedure lies somewhere between these extremes, following the latter in the broader classification of natural systems, but sometimes tending to the former in cases of small differences. It is hardly a matter of convention that an elephant is classified as different from a mouse (it does not make sense to say that a mouse is a kind of elephant), but there is something of convention in the distinction of a mammoth from an elephant (it does make sense to say that a mammoth was a kind of elephant).

The ideal of description, in terms of quantized observables and the mathematical analogy leads scientific workers to a continual attempt to express all qualities in predicates that are countables or measurables, and all descriptions in terms of relations between the predicates so expressed. I distinguish two phases in the remaking of predicates nearer to this ideal, the first I shall call, using Carnap's expression[3], *explication*, and the second, *reduction*.

The process of explication of a predicate consists of a change of expression of the predicate from qualitative to quantitative. There are two connected pressures which the process of explication satisfies; the desire for precision of expression of the observables which are used in giving a description of the state or change of state of a system, and the wish to find a functional generalization through the mathematical analogy that will express the facts. Clearly a number-variable cannot stand for an observable (the minimum condition of the mathematical analogy being applicable) unless the observable is expressed as a countable or measurable or both. The most that can be achieved towards the mathematical analogy and the expression of a general description in a function of observables without quantization is a generalization of the vague sort in which we express our intuitive apprehension of 'functional dependence'—'The more the merrier', 'The harder you hit it the faster it goes'. What can chiefly be objected to in these forms of expression is their vagueness, their lack of precision. 'Just how much faster will it go for a harder hit?' we ask and these intuitive generalizations will not give the answer. The need for precision then leads to explication, as we explicate the qualitative 'faster' into the quantitative 'increase in velocity of x ft. per sec.' And this in its turn facilitates the use of the mathematical analogy, for our raw material of varying observables is now expressible in the raw material of mathematical functions, the number-variables.

How is explication to be achieved? In at least two ways:

(i) By substitution of the unquantified property by a quantified form of the same property. This is what we have called an observable.

(ii) If this substitution cannot easily be made, then we can link the unquantified qualitative property with another property which is quantifiable, and so measurable. A condition of our being able to do this is that the property which we can measure should vary with the qualitive property which it is being used to explicate. If not exhaustive of the methods of explication these are at least typical procedures and will serve to illustrate the character of the process. The former I shall call *linear explication* and the latter *analogical explication*. The transition from speed to velocity is an example of linear explication, since we might say that the concept of velocity, quantitative and precisely determinable, is only a developed quantized form of the notion of speed. There is no abrupt transition to a new system of description when we substitute '. . . has a velocity of 50 ft./sec.' for '. . . has a high speed', or for '. . . is moving fairly quickly'. There is more to the explication of speed into velocity than this simple substitution relation, for there is the matter of the choice of units but that can be passed over for the moment since it is my purpose here to clarify only the difference between linear and analogical explications. The substitution of '. . . has a temperature of 65° C' for '. . . is fairly hot' is an analogical explication for it is not to substitute one predicate for another of the same kind since '. . . has a temperature of 65° C' is meaningless without the thermometer, while '. . . is fairly hot' is informative of how something feels. The justification of the substitution is through the connection that is found to exist between how hot or cold a thing feels and how much a column of a certain fluid expands. The superior reliability, that is their ability to be standardized, as well as the measurable character of temperature determinations determines the acceptance of the parallel between apparently quite disparate processes—the variation in a certain feeling, and the variation in the length of a column of a fluid. 'Hotness,' we say, 'has been explicated into temperature.' The one is a measure of the other.

Most properties can be explicated with a little ingenuity either linearly or analogically or both. The next phase in their re-expression as more desirable forms of description is obtained in another way, that which I have called above *reduction*. When the Greek investigators began the attempt to systematize and make precise the description of nature they set out to achieve this aim in the following way. It is said of Thales, the first Greek of whose scientifically minded investigations some record exists, that he attempted to reduce all things to water, and all properties of things to the states of water. Other Greeks attempted different reductions, a famous one which persisted down into the Middle Ages was the attempt to express all properties in terms of Hot, Cold, Wet and Dry. The details of this reduction are only of historical interest but in the method lies an important logical invention, the selection of one set of properties in terms of which others, and ideally all others, are to be expressed. The selection which the Middle Ages adopted from Greece was bad in that it led to no very powerful theory. However, the choice of spatio-temporal arrangement and mass of components of a system in which to express its properties, a choice made by the scientists of the Renaissance, led to modern methods of description. The properties selected were all measurables, and their connections were expressible in geometrical terms the only highly developed mathematics of the day. Dampier, discussing this new selection of basic descriptive properties, links up their selection with a new aim, that Galileo and others were content to aim at an accurate description of nature renouncing explanations. He says[4], 'The scholastic substances and causes, in terms of which motion had been loosely described in attempts to explain *why* things move were thus replaced by time, space, matter and force, concepts now first clearly defined and used mathematically to discover *how* things move, and to measure the actual velocities and accelerations of moving bodies.' Dampier is using 'why' here as the demand for a purpose in movement but the Renaissance development of accurate methods of description does not rule out why-questions in the other sense, as demands for satisfying explanations of happenings or classes of happenings. In fact as we shall see this kind of accurate description is an indispensable prerequisite for producing explanations which are in fact fully satisfactory.

The choice by Galileo and others made of basic properties in terms of which to express all other properties of things became incorporated into philosophic discussions as the distinction between primary and secondary qualities. Primary were those which it was thought were really properties of the object being described such as extension, quantity of matter and so on; secondary were those which depended for their character upon the state and orientation of an observer, such as colour, taste, feeling of temperature, and so on. It had been argued, though not shown in practice, by the Greek atomists that such a reduction was possible so that the ideas of Galileo were not new, but the ability to carry out experiments and actually to measure as quantities the qualities that had been selected as primary was indeed something that had hardly existed since the investigations of Archimedes.

Reduction of all qualities in the way described above has two connected aspects, in that it leads to a view, or perhaps springs from a view, of the structure of things as well as to a view of the 'structure' of observables. Galileo's measurements of distances, times and masses had a rationale in the belief he held that things were made up of other things whose important properties were their spatial arangements at various times and the quantity of matter they contained. The connection between the rationale and the choice of basic properties is very close, so that from a logical point of view the study of reduction of qualities is at the same time an exhibition of the prevailing view of the structure of things. A typical reduction of Galileo's time was that of velocity. In medieval treatises it could be treated as a quality of a body along with others like colour, size and weight. Its reduction to a function of space traversed and time taken makes it not a genuine, irreduceable, intrinsic property, but a construction out of well-defined measurables. In mechanics and general physics the success of reduction had reached such a stage by the nineteeth century that it was possible to say that in theory there were just three physical 'dimensions' in terms of which every physical observable, every quality of bodies that could be included in a physical description, can be expressed, and to constructions of which they could be reduced. In practice, however, these dimensions were and are used for the expression of units of measurement, rather than for the design of experimental equipment. The dimensions are extension, [L], time, [T], and mass, [M]. It was popular at one time among philosophers to read back this reduction from observables into the world and to speak of nature as a complex of 'mass-points in a spatio-temporal framework'.

In the early part of this chapter I described the basic principles of quantitative description in terms of a unit defined by a certain set of physical operations and of a measurement, which is the cardinal number of unit operations required to complete a certain measuring procedure. What physical operations are chosen as defining the units of measurement will depend very much on the reductions that are acceptable. This move, which was called 'operational definition' by Professor P. W. Bridgman, is a second stage in the reduction of observables beyond that attempted by Galileo. Bridgman[5] discusses the reduction of the concept of length, one of the basic observables or primary qualities into terms of which Galileo's first stage reductions were made. This has itself, Bridgman argues, a dependence upon things other than the object being measured which makes it, until further reduced, ambiguous and vague. Those things upon which it depends for precise significance in an experimental-situation are the operations by means of which length is measured. It is necessary then to make a further, and, Bridgman believes, final reduction of observables, and to break up for example the observable 'length' into a construction of operations of measurement, say the laying of yardsticks along an object according to a certain procedure. Bridgman goes further and claims that this and other constructions are what 'length' actually means in various branches of physics, but we do not need to adopt this extreme view of the meaning of observables to see the importance of this second stage of reduction. By the two stages of reduction now described the quantitative determination of any observable can be made absolutely unambiguous and its value, within the limits of the experimental techniques available, precise. Ambiguity is avoided by expressing it in terms of the chosen primary qualities, expressed in the choice of physical dimensions (for example 'Velocity is the quotient of distance travelled and time taken and has units $[L]/[T]$'). Precision is achieved by expressing the dimensions as constructions of operations, and so defining the units (1 yard is a certain combination of sliding and marking operations with a yard-stick). In this way quantitative and unambiguous determinations of the primary observables can be made in the way suggested by the dimensional analysis of the unit of velocity '$[L]/[T]$'.

For many qualities reduction is not so direct and simple a process. Heat, for example, can be given an analogical explication into temperature, but to make any further move a reduction of temperature must be made. I suggested above that the actual direction of Galileo's reductions showed a certain view of the structure of the matter. To make any reduction of a notion like temperature requires just such a view of the structure of matter, for we have to be able to say what, in terms of this view the temperature of something is, before we can attempt the reduction of the quality in terms of primaries and the definition of a unit in terms of the physical dimensions. In fact, if anything more than the explication of qualities is required, if the second phase in the making of predicates is to be attempted, a general view of the constitution of nature is involved. Reduction then is closely connected with the accepted theory in the field being investigated, and with the way that an explanation of the qualities to be reduced is given. Reduction, particularly of the important class of dispositional predicates, can only be understood when the role and function of theory has been studied.

The methods of description that have been outlined are only one aspect of the whole process of giving a scientific description. Though a haphazard collection of descriptive generalizations may in fact be entirely comprehensive and each one a very precise description, if the collection lacks organization, it is not shown by their practice to be what scientists want. Indeed they are often prepared to sacrifice something of precision and much of comprehensiveness if some system and order can be imported into a set of generalizations. There seem to be two ways in which the knowledge that is expressed in generalizations has been systematized—the first I shall call *Aristotelian systematization*, and the second *Galilean systematization*, since Aristotle and Galileo, though not the originators of the respective methods, popularized them.

I. *Aristotelian Systematization*. Knowledge is ordered by ordering the objects, the things which we know something about. The knowledge that we express in generalizations is turned back upon the world and used to make a classification of these things. The procedure of classification can be imagined first as a practical sorting of the things in the field that has been chosen for study.

A hunter brings the results of his expeditions to a taxonomist. Suppose that the taxonomist has given an order that he shall hunt only animals. The hunter brings back the carcases of innumerable creatures and piles them on the taxonomist's floor. The taxonomist sorts them into convenient bins. Vertebrates he puts in one bin, invertebrates in another. Then he sorts the vertebrates in mammals, fishes, birds and so on; and the invertebrates into coelenterata, insects, and so on. He resorts these sub-bins until he has in the end only animals of 'exactly' the same appearance and characteristics in each separate bin. 'Exactly' means here what the taxonomist wants it to mean, that is there may be differences which he is prepared to overlook, *e.g.* in weight provided that the differences are not too great. The process described here in practical terms can be given two different logical descriptions. We can take it either as proceeding from a 'rough' classification depending upon a single feature, say the presence or absence of a backbone, through a series of further classifications according to single points of difference until no further points of difference can be found which are judged important: or as a classification in the opposite direction, sorting according to a unique *set* of features which distinguish one animal from every other. By progressively reducing the number of features which form the set according to which classification is made, the same end result can be reached. It should be observed that in the latter case it is bins of animals which are progressively reclassified.

In the technical language of taxonomists the processes described above lead to the differentiation of kingdoms, phyla, classes, orders, families, genera and species. Procedure in the direction of kingdom to species as in the first sorting method above I call systematization by classification, in which the differentia of a species, the characteristics which uniquely define it, are built up by the addition of a further characteristic at each resorting. Procedure in the direction of species to kingdom as in the second method above I call systematization by abstraction, for at each resorting the number of characteristics which distinguish one bin from another are progressively reduced. The set of characteristics which determine which creatures fall into a given group is, in the jargon of logic, the *intension* of the group or class it specifies, while the members so grouped are the *extension*. The main features of Aristotelian systematization can be summed up in the classical slogan 'Extension varies inversely as intension', that is the fewer the features of difference that are remarked upon the more members a class marked out by these features will contain. Aristotelian systematization is clearly of great importance in the biological sciences and indeed it can be argued that Aristotle who popularized this method of systematizing information derived it from the necessities of his own biological studies.

There is nothing definite or final about the results of this method of systematization for any Aristotelian hierarchy of classes that is built up depends for its particular form upon which principle of classification the builder has in mind. The present classifications of biological individuals are very different from the earlier attempts, the difference depending partly upon the taxonomist having more information, and partly, though this is connected with his having more information, upon the use of a different principle of classification. At one time plants were classified according to their medicinal properties, then various structural features were selected such as the number of rudimentary leaves in the embryo, and now as an extension of this, biological species are classified according to their evolutionary groupings. Three different principles of classification have been used, and the fact that the latter pair give a closely similar result shows how closely structural similarities and theories of evolutionary history are connected.

Attempts were made by the early scientists to apply the Aristotelian systematization to every field of knowledge, a commonly accepted principle of classification being the political. Each phylum of species was imagined to have a 'feudal' organization, the whole world forming a vast system of systems. That the lion is King of the Beasts is a nursery survival of what was at one time a general system of classification. Minerals too were similarly organized with gold the royal metal and diamond the royal earth. It is easy to see from this example that a principle of classification cannot be entirely arbitrary, for the uselessness of this classification reflects the fact that there is no political or quasi-political connection between substances or even among species for that matter. Political organization is not even a close analogue of the differences between substances and the system had to be abandoned.

II. *Galilean Systematization.* Knowledge, in the Aristotelian method, was systematized by ordering the objects of knowledge, and by using generalizations simply to facilitate the ordering. A more direct way to bring order into our knowledge is to try to systematize the knowledge expressed in a set of generalizations by ordering the generalizations. One method of so ordering them that has proved to be so useful as to have been almost universally adopted is to organize the generalizations to hand into a deductive system. This kind of ordering is greatly facilitated by mathematical expression of the generalizations for one way we can use the results of pure mathematics is as rules for the manipulation of symbols used in functional expressions. Galileo, though not the first to use the method of deductive ordering, succeeded in giving such clear mathematical expression to the generalizations he discovered that systematization could be carried out with the whole apparatus of mathematics. Newton's general mechanics of the *Principia* was the first really comprehensive system of this kind and has provided the model for all subsequent attempts. There were considerable difficulties in developing the most useful language for the expression of generalizations, the mathematical function expressing the dependence of variables one upon another. Newton by providing the necessary mathematical instrument was able to systematize the laws of rectilinear motion into a deductive system. Suppose we begin with a condition, that acceleration is constant in the motion, which can be expressed as a differential equation:

$$\frac{\mathrm{d}^2 s}{\mathrm{d}t^2} = a \tag{1}$$

One of the sets of mathematical rules of inference can be applied to this condition and we get:

$$\int \frac{\mathrm{d}^2 s}{\mathrm{d}t^2} = \int a\, \mathrm{d}t + K$$

$$\frac{ds}{dt} = \quad at + K \tag{2}$$

K is a constant, which is an expression of the conditions at the beginning of the motion when $t = 0$ and is the initial velocity, say u. ds/dt is the differential expression for velocity v. We have deduced a law of motion:

$$v = at + u \tag{3}$$

which could be verified directly. Applying the rules of integration to (2) we get:

$$\int \frac{ds}{dt} = \int at \, dt + \int K \, dt + L$$

$$s = \tfrac{1}{2}at^2 + \quad Kt + L$$

Since at the beginning of the motion both distance s and time t are zero, it follows by simple algebra that L disappears. K we know to be the initial velocity u. Another law of motion has been deduced:

$$s = \tfrac{1}{2}at^2 + \quad ut \tag{4}$$

which could again be verified directly from the measurements of the appropriate observables in an actual motion. This example shows in miniature the main features of the methods of systematization that are now used in the physical sciences.

Mathematical expression, though extremely convenient is not a necessary condition for the development of deductive systems. Generalizations can also be systematized into deductive chains when they are expressed in ordinary language and when the rules of inference are the rules of ordinary logic. An example of such systematization was discussed in detail in Chapter One (p. 32). The following taken from a work by R. E. Peierls[6] is a closely reasoned, though minute deductive fragment. 'We had already seen that chemistry required attractive forces between atoms in some cases, but also that this attraction would change into repulsion when the atoms actually came into contact with each other. If then I have a large number of atoms in a small space the forces try to get the atoms as close together as possible without their actually overlapping, and this leads in general to some rather regular pattern.' Briefly analysed this passage has the structure of a deductive system:

(1) ... Many atoms in a small space are close together.

so by (I)

(2) They tend to get closer together.

so by (II)

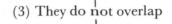

(3) They do not overlap

so by (III)

(4) They tend to form regular patterns.

'Rule' I. There are attractive forces between atoms.

'Rule' II. Atoms in contact repel each other.

'Rule' III. Atoms which are tending together but do not overlap tend to form regular patterns.

A feature of this example that is worth noting, is that a passage of deductivity related statements may contain the 'rules' of inference according to which the conclusions are drawn as well as the statements forming the premises and conclusions of the deduction. These 'rules' themselves may be deductively related. This is always the case when the conclusions are drawn according to the rules expressed in a mathematical formalism, and is the case too with the informally organized passage quoted above. I am sure that it would be agreed by scientists working on the structure of matter that either they have succeeded or hope to succeed in expressing the laws one of whose applications is as the rules I and II above ('There are attractive forces betweeen non-contacting atoms' and 'There are repulsive forces between contacting atoms') as consequences or parts of a single deductive system, the general theory of the structure of matter. 'Atoms which are attracted at short distances from one another and are repelled when they come into contact tend to form regular patterns' is really a statement of a general law of topology applied to a particular case, and it too should be deducible within its appropriate deductive systems. Organized descriptions do not in general form simple deductive systems, but rather consist of a number of such systems more or less simply connected through the conclusions of some being the basis of the rules of inference for others.

An account of scientific description would be incomplete without some discussion of the way particular facts are related to the generalized descriptions that it has been my main purpose to analyse. I do not wish in this part to discuss the questions which arise from the relations between facts and generalization as the facts confirm, give grounds, reasons or support for generalizations. These questions are included in the topics of induction and confirmation. But I do want to make clear here the way generalizations can be used to make detailed predictions, how they facilitate the inferences that distinguish prediction from guessing. This will introduce another basic function of science— prediction. This function can be treated only in a very limited way at this stage of the investigation for theory provides techniques of prediction that are more powerful and comprehensive than those description-based techniques which I shall describe here.

A mistake that is frequently made in describing prediction-inferences is to suppose that by a continual, logically smooth process transition can be made from the general to the particular. In fact we can never say what is the case in a particular instance from a generalization alone, but only from a generalization in combination with a particular providing, as we might say, location for the prediction of the instance. The process of prediction with the help of generalizations can be variously expressed. The commonest forms used by logicians have been traditionally the syllogism and the hypothetical argument of which a characteristic form is the argument *modus ponens*. Examples of these, chosen deliberately for their banality are:

(*a*) *The syllogism.*
        (i) All rabbits are herbivorous.
        (ii) This is a rabbit.
Therefore (iii) This creature is herbivorous.

(*b*) *Modus Ponens.*
        (i) If a creature is a rabbit then it is herbivorous.
        (ii) This is a rabbit.
Therefore (iii) This creature is herbivorous.

These forms illustrate very clearly the traditional but mistaken view of prediction for it has been supposed that the general statements a(i) and b(i) above can be always treated as premises in the arguments to which they are relevant; that is to suppose that the enunciation of the argument forms above settled the elucidation of the inference to a particular case. But if we pause for a moment to consider the way instantiation is actually performed we can see that purely logical considerations apart the schematizations above do not truly represent the process. The natural process of prediction of an instance is to state the instance as a consequence of another instance, for example that the creature is herbivorous follows from the fact that it's a rabbit. The justification of this move, marked in ordinary language by 'so' or 'therefore', etc., takes us back to a generalization or its corresponding conditional, that is to a statement of the form of either a(i) or b(i). These are not premises since they validate but do not belong in the argument that expresses the deduction. It is natural to call them the rules of the deduction. We infer a particular not from a generalization but in accordance with it. This becomes very much clearer in those cases where the generalization is a functional expression and the particulars are stated in terms of the numerically determined observables. Suppose for example that a stone is dropped from a tower; to compute (infer by mathematical rules) the distance fallen in, say, two seconds the following procedure is followed:

(i) The particular facts which form the basis of the inference are set out; (t) is 2 sec., initial velocity (u) is 0 mps, acceleration is that due to gravity (g), 9.8 mps.

(ii) A calculation of the unknown particular, distance fallen (s), is made by using the generalization $s = ut + \frac{1}{2}gt^2$, not as another premise along with the facts in (i) but as rule of inference, the rule that allows the calculation of s. The detail is as follows:

$$s = 0 \times 2 + \tfrac{1}{2} \times 9.8 \times 4$$

$$s = 19.6 \text{ m}.$$

The logical point, that in making predictions the rule of inference does not form one of the premises, can be reinforced by an argument from Lewis Carroll[7]. Consider the inference 'p therefore q'. We may justify the deduction by insisting on the truth of 'If p then q'. If it is now insisted that since this statement is relevant to the conclusion it should be included in the premises of the argument then it can be shown that no argument can be represented with less than an infinity of premises. In its expanded form, according to the relevancy requirement, the argument becomes 'If p then q, and p; therefore q'. The justification of this 'therefore' is the rule 'If, if p then q and p then q'. Insistence on the inclusion of statements whose truth is relevant to the conclusion would lead to the inclusion of this statement among the premises. Clearly this is the beginning of a regress of requirements which can have no end. The way out of the difficulty is simple; we must not be beguiled into taking even the first step, for true statements can be relevant to a conclusion either as premises in an argument that leads to the conclusion or as rules according to which the conclusions of the argument are drawn. There is no problem if rules of inference and premises of inference are kept apart. In the deduction of particulars that is the basis of prediction, the premises are particular and the rules of inference are general. It is for this reaon that I should like to say that science is the *logic of nature*, for it consists of all those generalizations which can be appealed to in justification of the deduction of new facts from old. This point provides an additional reason for the use of numerically expressed measurements for the representation of observables; namely that via functional generalizations algebraically expressed the most powerful rules for making predictions can be developed.

In this chapter the basic logic of two of the central functions of science has been investigated, description and prediction. It has turned out to be the case that in the process of generalization and systematization we have the necessary logic machinery to form the basis of both functions. There remains the function of explanation.

REFERENCES

1. H. Margenau, *The Nature of Physical Reality*, Chap. 8, Sect. 8.2. p. 171.
2. P. F. Strawson, *Introduction to Logical Theory*, pp. 239, 240.
3. R. Carnap, *Logical Foundations of Probability*, Chap. I.
4. Sir W. C. Dampier, *A History of Science*, p. 146.
5. P. W. Bridgman, *The Logic of Modern Physics*, pp. 9 ff.
6. R. Peierls, *The Laws of Nature*, p. 114.
7. L. Carroll, 'What the Tortoise said to Achilles', *Mind*, N.S., **4**, p. 278 (1895).

4 Scientific Explanation

The preliminary analysis of explanation made in Chapter 1 brought out two points of great importance. The first was that a particular happening is explained when the conditions under which the happening occurs are isolated. It was found that a logical condition for this isolation was the existence of a generalization or generalizations linking the happening to be explained with other happenings which could be settled on as a whole or part cause of the happening. It must be emphasized that this procedure provides explanations only of particular happenings. When it becomes necessary to give an explanation of types of happenings, of classes of events, explanation cannot be confined to isolating particular causal conditions, but, and this is the second point brought out in the general discussion in Chapter 1, an explanation must enable us to understand. And understanding is gained either by our finding an illuminating analogy to the phenomena whose character we do not understand, or by our 'exposing a hidden mechanism' the workings of which inevitably result in the phenomena that required explanation. An explanatory theory may depend for its acceptance on the success of its analogies, or on the plausibility of the mechanism it postulates, or in many cases on both. As we shall see, at least in scientific explanation, the building up of analogies and the investigation of the fine structure of the world interact.

The first type of explanation, I would like to discuss is not greatly different from the systematized descriptions of the last chapter. However, it was characteristic of those descriptions that they were, logically, all on one level. The mathematically related kinematic laws linked up observables directly without introducing concepts of which the observables, or rather the particular values of observables would be a consequence. In giving an explanation, the explanatory material must be on a different logical level from what it is to be explained, which must be seen, in one of many different ways, as a consequence of it. I would like to discuss an example from ecology, in which one set of facts descriptive of some phenomenon *A*, is brought in to explain another set of facts descriptive of some phenomenon *B*. A connection is established in the theory between *A* and *B*, this connection being the necessary condition of the description of *A* being an explanation of *B*.

'Our knowledge of the conditions of existence in deep oceanic waters is still rather limited, but recent studies of the Sargasso Sea have given us a clearer conception. It seems that the annual production of living matter by the microscopic plants of this supposedly barren sea is not so much less than in neighbouring shallow waters. There is, however, an important difference in that the vertical spread of the phytoplankton is greater in the deep sea waters, meaning that the number of plants per unit volume of sea water is fewer than in the shallow coastal seas where the plant populations are concentrated in a smaller space. So a herbivorous oceanic animal must work harder to obtain the same amount of plant food than does a shallow water relative with the same food collecting capacities. To build up a given amount of new living tissue the deep-sea herbivore must expend in energy a greater proportion of the food it has collected. In other words a given volume of the deep ocean will be less productive of herbivorous animals, such as copepods, than a given volume of the shallow seas, and this in turn will mean a reduced production of the carnivorous species. (But there is another aspect to be kept in mind. An animal feeding on the denser plant crops of the shallow seas may not digest all the food it collects, whereas the deep-water animal may extract the last particle of nourishment from its harder-won food.)

'Now these differing productivities seem to be reflected in the structure and habits of shallow-sea and deep-sea plankton-feeding fishes. In general the herring-like fishes (Clupedae) of the shallow seas screen out the copepods and other small forms of life on which they depend by means of a fine meshwork of gill-rakers set across the gill-arches. Filter-feeding fishes are not found in the depths of the ocean. The lantern-fishes and certain of the stomiatoid fishes which feed on small planktonic animals may have numerous gill-rakers, but these never form a fine sieve such as is found in the herring. Most probably these deep water fishes snap up their food as they come upon it.'[1]

This passage aims at explaining the fact that the fish found in shallow water have a fine screen of gill-rakers suitable for the filtering out of food-material but the fish found in the deep sea do not. Two sets of facts combine to provide an explanation; the first, S_1, the distribution and productivity of plant life in various parts of the sea; the second, S_2, the uses that are made of food and food-energy by herbivorous and carnivorous fishes. From each set conclusions about oceanic productivity can be drawn.

C_1. From S_1: deep seas have a lower concentration of plants.

C_2. From S_2 and C_1: there will be a smaller weight of herbivorous fish, as food for carnivores, in the deep sea than in shallow waters.

C_3. From C_2 and the use of R, 'Filter feeding is effective in densely populated areas and hunting in thinly populated' as a rule of inference: the fish of shallow seas will be equipped with a fine mesh of gill-rakers for filtering out food, and deep sea fish will not.

In this way the structural differences that required explanation are seen as the 'outcome' of certain facts about the sea, together with a simplified account of the use that an animal makes of its food. 'Outcome' here is not intended strictly to refer to a result or effect, since all that is required for an explanation to be minimally acceptable is logical connection. As we saw in Chapter 1, a cause is always a selection from those facts which are judged relevant. By means of the generalization R, connection is established between feeding methods and food density, and through this connection the relevancy of the facts about distribution of oceanic plants to the differences in feeding habits of fish is made clear. Feeding structures and feeding habits are clearly closely connected. 'Clearly' used here simply expresses our familiarity with the background of general biological beliefs of which the adaptation to environment is a part. Through the multiple connections the relevancy of plant distribution to structural differences is established and the one serves as an explanation of the other. The logical character of this kind of explanation, apart from the complexity of the connections establishing the relevancy of certain facts to one another, is the same as that of Type IIIa, the general explanation in detail of Chapter 1. The characteristics of a species, in this case the feeding structures, are explained by some class of facts about the environment whose relevancy is established by suitable generalizations.

The connections which establish the relevancy of classes of facts to one another are often extremely complex, so that if they are to be of any use in facilitating understanding, which is a requirement of an explanation over and above delimiting the area in which we can find a cause, some simplification of the relevant facts must be made. The most direct way, and that commonly adopted in biology is to represent a structure or a process or an environment schematically, so that only the most important features are brought out. There is something of this in the extract above for:

(i) The oceans are treated as if they had only two distinguishable regions, a deep-sea and a shallow-sea, each with a characteristic flora and fauna.

(ii) The life of a fish is broken down into food-gathering and body-building activities; and the food collected thought of as partitioned out according to the demands of these activities. I should like to call this process of schematization, *forming a picture of the facts*. An explanation provides a picture of the facts. When this picture has been grasped then an understanding has been gained. It is this feature that distinguishes this kind of explanation from the deductively related system of descriptive generalizations of the last chapter, for from those, causal statements could also be derived.

At this point it will be useful to introduce the notion of a *model*. A model, *a*, of a thing, *A*, is, in one of many possible ways, a replica or an analogue of *A*. The most familiar kind of model, is the scale-model, where an exact copy of something is made, but of a reduced size. Models of this kind I shall call in general, *micromorphs*. The model, *a*, is related to its 'parent', *A*, in some quite definite way which could be expressed in rules according to which, in the simplest case, the dimensions of the one could be turned into the dimensions of the other. Micromorphs play an important part in engineering for it may be more convenient to study a model than the real thing. Familiar examples are the little aeroplanes that are suspended in wind-tunnels, and the models of the sea-floor that can be used, by means of a miniature and greatly accelerated ebb and flow of the tides, to predict the conformation of a harbour in many years time. A simple scaling up of the results obtained in miniature is often not successful in translating the information found by the micromorph into a form applicable to the full-scale object, so that elaborate experiments are sometimes necessary to discover just what are the correct rules of transformation from micromorph to original.

Micromorphs are confined essentially to the modelling of things, and can only serve to represent a process when the process itself is run on a small scale in micromorphs of the gear required for running it on a normal scale. Pilot plants are usually of this type. It very often happens, however, that there is a parallel between the form of the laws in two otherwise quite different fields of investigation. Such parallels can be exploited in the development of another kind of model, in which processes can be represented in some analogous system. I shall call this kind of model a *paramorph*. Just as for micromorphs, so for paramorphs there are rules for transforming information obtained in the paramorph into information about the parent, and these rules may be exceedingly complicated, and require careful experimentation before they can be exactly settled. Paramorphs have important experimental applications. For example there is a close analogy between the laws of transfer of mass, momentum and heat across various kinds of boundaries. This analogy can be put to work to solve problems of great practical difficulty, *e.g.* the investigation of heat transfer to the circulating liquid in a tube-boiler is a difficult problem in practice. The analogy between mass-transfer and heat-transfer is exploited in the construction of a paramorph in which rods appropriately coated are allowed to dissolve in the circulating liquid so providing a parallel to the transfer of heat. The rate of loss of mass can be very accurately determined and by the use of previously determined rules of transformation can be transformed into information about heat-transfer.

The analogy that is found to hold between certain characteristics of different processes is the basis of the paramorph. In the example I discussed above the paramorph was actually constructed, used and experimented upon, but the construction need not actually have been carried out to provide our thinking with a powerful support. In this way we make for ourselves conceptual models, in which something with which we are familiar or which we understand very well is used as an imaginary model of some otherwise obscure process. The analogy is the simplest kind of conceptual paramorph. Now just as analogies were found to be a basic element in the construction of the ordinary explanations analysed in Chapter 1, so the more general notion of a conceptual paramorph is a basic element in the analysis of the construction of scientific explanations. One might go so far as to say that in many cases an explanation of phenomena *A* is provided by describing *B*, a conceptual paramorph of *A*. The full-scale theory based upon a paramorph must then contain three logically distinct parts:

(i) The description of A, the facts requiring explanation.

(ii) A description of B, the appropriate conceptual paramorph.

(iii) A series of transformation statements linking B to A, statements determining the relevancy of B to A.

This analysis of a kind of scientific explanation has been made in different but equally elegant ways by both Campbell[2] and Smart[3], each contrasting a formal description systematized as in the preceding chapter with a theory which provides an explanation by linking a model with the facts to be explained. Of course the conceptual paramorph which is selected for use in the theory under construction can be described either informally or formally, that is in a pictorial way without the use of mathematics, or by means of a deductive system of functional generalizations, or both in different measures. I should like to analyse an example typical of each mode of description, for in practice no theory is so simple as to fit exactly the analysis (i), (ii), (iii) above.

A neat, relatively self-contained example of the use of an informally described model is provided by Fajan's theory which provides an explanation of certain apparent anomalies in the chemical behaviour of elements arranged in the classical periodic table. The phenomena to be accounted for by the theory are:

(i) Beryllium chloride and similar salts are poor conductors of electricity.

(ii) Sodium chloride and similar salts are good conductors.

The behaviour of beryllium chloride is apparently anomalous since one would expect from the behaviour of sodium chloride that all chlorides would be good conductors. Some differences in the chemical behaviour of sodium and beryllium that may be relevant are; sodium occurs towards the beginning of the group with which it is classified in the table, while beryllium occurs towards the middle; elements similarly placed to beryllium tend to form covalent compounds (that is compounds which do not readily split into charged ions) while elements similarly placed to sodium tend to be electrovalent (that is forms ions readily). Fajan's explanation of the anomalous behaviour of beryllium compounds runs as follows: 'In a molecule consisting of two ions the electrostatic attraction holding the ions together will tend to draw them closer, with the result that the "orbits" of the outer electrons become distorted or deformed. If this attraction becomes sufficiently great the distortion will become so considerable that some of the electrons instead of being associated with one of the ions will be shared between both. That is to say electro-valency will tend to pass over into co-valency when the attraction between the ions is large'[4]. In a word because beryllium compounds, especially the chloride, are characterized by a strong attraction between the ionic parts of the compound molecules they will really be covalent, and so it should not be expected that they should conduct electricity readily.

This theory may be developed to suggest, and predict the results of, new experiments. It can be argued that because the attraction between ions will be greater when their electrostatic charge is greater, ions with a high valency, that is with a large unbalanced charge, will suffer more distortion and so tend to form covalent rather than electrovalent compounds. The experimental consequences of this would be that compounds formed from elements which occur towards the middle of a series in the periodic table are poor electrical conductors. This is in fact confirmed by experiment. Further it can be argued that because the electrostatic field is greater round a small charged body and weaker round a large one, compounds formed from ions in which one of the ions is large and the other small will tend to be covalent. The experimental consequence of this, that compounds formed from a pair of ions of which one is an ion from an element which is classified towards the beginning of the periodic table and the other towards the end, will be poor conductors of electricity, is confirmed. There are other deductions which can be made from the theory but these will do for the present purpose.

Our theory, or explanation, is first and foremost, an electro-chemical explanation, that is it is within the current general theory of the structure of matter and so already presupposes a model. We have become so used to the correlation of overt happenings with modifications to the character and arrangement of electrically charged particles that it is hard for a moment to grasp that this elaborate conceptual structure is based in fact upon a very comprehensive model, built up on the overt principles of electrostatics. There is now good reason to suppose that the description of this model is also a fairly correct, though rough and ready, description of the fine structure of things. The facts of electrostatics, that for example, similarly charged bodies repel each other, are put to work as the principles of construction of the paramorph that underlies the electro-chemical view of matter. Couple with this general theory of the structure of matter the planetary model of the atom, the basic unit of the electro-chemical view, and we have the framework in which Fajan's theory can be enunciated. The innovation which it introduces is the idea of deformation, breaking and reforming of electron orbits, which, though an extension, is a natural extension of the conceptual paramorph within which it is framed. Fajan's theory, then, superficially so simple, depends on the interaction of three models, each a modification and extension of the last. The theory so built is systematic, at least to some extent, and deductions can be made in it, but it remains inexact, pictorial and vague.

In discussing scientific description I made the point that when great precision of description is the aim, then the observables selected are quantifiable, that is either measurables or countables and the expression of a generalization is in terms of a function of variables. Now in developing a general description a variable represents a range of definite values, that is it represents a certain observable unspecifically. This being so the function/variable mode of expression is eminently suitable for describing the conceptual models which are the basis of explanations, since a variable can represent the unspecified and imaginary observables of a conceptual model too. I should make it clear at this point that I am discussing the construction of theory as if models were invented spontaneously having no ground in the facts, and as if the question of their reality never arose. This order may not be the actual order in which any theory was built, but I shall preserve the logical fiction that models are first invented, then described and linked to the reality they parallel and finally tested for whether or not the invention has hit off the fine structure of the phenomena explained.

As an example of a theory in which a model is given a function/variable description I shall take the scientifically outmoded but logically very clear case of Bohr's original theory of atomic spectra[5]. He pictures matter as made up of two fundamental kinds of particle, one of positive charge, and the other negative. The number of charged particles making up an atom is related to experimentally determined observables by the postulates:

(i) The positive charge carried by every atom is numerically equal to the atomic number Z.

(ii) The number of electrons retained by an atom is also equal to the atomic number Z.

Bohr's mathematical expressions derive from function/variable descriptions of Rutherford's planetary model of the atom, that is the positive charge and most of the mass is supposed to reside in a heavy nucleus around which move planetary electrons of a number sufficient to balance the charge on the nucleus. According to this model the equilibrium of an orbital electron will depend on the equality of centrifugal mechanical forces and centripetal electric forces. If an electron has a mass m, a charge $-e$, and is moving with velocity v in a circular orbit of radius r, around a nucleus of charge $+e$, then the electron will remain in dynamic equilibrium if

$$\frac{mv^2}{r} = \frac{e^2}{r^2}$$

This equation will not account for the fact that light is emitted in quite definite wave-lengths from substances, and for this reason the model must be modified by a pair of *ad hoc* hypotheses.

(i) Electrons can move only in orbits of certain definite radii according to the rule $2\pi \times$ angular momentum (mvr) is equal to nh (where n is an integer and h is Planck's radiation constant).

(ii) Light is emitted from a substance only when an electron in an atom of the substance jumps instantaneously, without spiralling, from one orbit to another.

Making use of these hypotheses it is clear that if the energy at level l' and E' and at l'' is E'' then the energy of the emitted light will be

$$E' - E''$$

which according to Planck is equal to $h\nu$, where h is a constant and ν the wave-length of the light. Knowing the various possible orbits, and so the various possible energy differences, it is possible to calculate quite easily the possible wave-lengths of emitted light, for the simplest atom, that of hydrogen. The wave-lengths found in this way are experimentally recognizable in the so-called Balmer series of emission lines in the spectrum of hydrogen.

Bohr seems here to have constructed something superficially very like the deductively systematized description of the last chapter, for by the substitution of definite values for variables the value of an observable like wave-length can be calculated. The difference lies in this, that while every variable in the kinematic equations stood for an observable, and the kinematic equations were directly descriptive of those motions for whose particular observables, they were the rules of predictive inference; in the theory just considered the algebraic functions are not directly descriptive of the observables of an actual situation but describe the chosen model. Only some consequences of the model's behaviour appear in the macroscopic world. Only some properties that such a mechanism could have are judged to be relevant (*e.g.* we do not enquire of what material it is made, what its colour is, etc.) so that, whether they are explicitly formulated or not there must exist rules for transforming properties of the model into observables. The facts predicted by Bohr from his atomic theory are vastly different in kind from the features of the model with which they begin, but the transformation is not arbitrary. The transformation rule

$$E' - E'' = h\nu$$

by means of which we pass from states of the atomic model represented by E', etc. to observables, the value of ν, has a rationale in the way we nowadays understand light in terms of energy. This understanding itself involves the interaction of many models, a development which has been described and analysed by S. Toulmin[6].

This method of theory construction, so common in physics and chemistry can be schematized as follows:

1. The laying down of principles in which the character of the model is outlined.

1. An atom is made up of a heavy nucleus with planetary electrons.

2. The building up of descriptions of the model (functional in Bohr's theory, informal in Fajan's).

2. *E.g.* 'Centrifugal mechanical force is equal to centripetal electrical attraction', i.e.

$$\frac{mv^2}{r} = \frac{e^2}{r^2}$$

3. The transformation of some or all of the model-describing expressions into descriptions of actual natural phenomena by means of rules of transformation, which must not be arbitrary but will be based upon the general view of nature (the overall model which is currently accepted).

3. *a. Description of model.* The electron in a hydrogen atom can move only in orbits determined by the relation

$$2\pi mvr = nh$$

b. Rule of transformation. Energy differences between the allowed orbits determine the wave-length of the light emitted.

$$E' - E'' = h\nu$$

> *c. Description of nature.* The spec-
> trum of incandescent hydrogen
> consists of a series of lines of definite
> wave-lengths.

It should be noticed that the theory may be extended either by deduction from the statements already descriptive of the model or by describing further features of the model. In the latter case it is not something objective that is being described. What further features we allow our model to possess is really a matter for our own convenience though no doubt dependent on the direction to which the facts point. The results of either method of extension will be more model-descriptions capable perhaps of transformation into descriptions of nature as in (3) above.

A further development in theory construction, foreseen to some extent by P. Duhem[7], has recently taken place. In certain branches of physics the facts which have been discovered have been of such a kind that until the present no satisfactory model has been hit upon such that these facts can be produced by transformations from it. A number of more or less loosely conceived models have been used from time to time with varying success but always failing to provide a satisfactory overall picture. The solution that has gradually come to be adopted is, in effect, to treat the mathematical formalism which in the example just discussed was descriptive of the model as the model itself. This of course involves giving up the hope of using the model as a guide to the character of ultimate reality (which in most cases means the fine structure of whatever it is that is being investigated), for it is just a muddle to say as Jeans sometimes does[8] that the ultimate character of the world has been shown to be mathematical. The world is the world, only our models can be mathematical. Examples of theories which could be said to depend upon mathematical models are to be found in the study of the fine structure of matter, for example the quantum mechanics formulated by P. A. M. Dirac. The exposition of such a theory is much too complex for my purposes here. Instead I shall try to point out the main logical features of taking this further logical step, without attempting to develop the theory itself in any detail.

The need for a mathematical construction as a model is explained by Dirac as follows: 'In the case of atomic phenomena no picture can be expected to exist in the usual sense of the word "picture", by which is meant a model functioning on essentially classical lines. One may, however, extend the meaning of the word "picture" to include any *way of looking at the fundamental laws which makes their self-consistency obvious.*' And this requirement for Dirac is met by expressing them in a mathematical scheme, that is to say the descriptions of nature in which observables occur have to be interpreted into a mathematical scheme, which if developed from consistent axioms and rules will itself be fully consistent.

But how does this sort of construction differ from the deductively organized, mathematically expressed systems of description such as the Galilean kinematics? In Bohr's theory above the difference was clear, that the theory described the model it had adopted as if it were real. The only difference of importance in Dirac's mechanics is that the mathematical system is a good deal wider than the descriptive generalizations which can be derived from it. In a way of course the laws of elementary kinematics have a mathematical model for it is only because the linear function of variables, say $v = u + at$, gives a table of definite values the same as that given by an experiment with an accelerating body that the function is chosen as the symbolic description of the motion. This was the underlying analogy which was noticed at the beginning of Chapter 2. The elementary kinematics is a limiting case, for there are neither mathematical nor observational loose ends. In quantum mechanics the analogy does not force itself upon us, nor can it be derived by abstraction from the results of experiments. It has to be deliberately sought. Dirac says, 'Quantum mechanics . . . requires the states of a dynamic system and the dynamic variables [observables] to be interconnected in quite strange ways that are unintelligible from the classical standpoint. The state and dynamic variables have to be represented by mathematical quantities of different natures from those ordinarily used in physics. [Thus ruling out the use of macroscopic processes as models, for it is from macroscopic processes that the variables of classical mechanics get their meaning and their life.] The new scheme becomes a precise physical theory when all the axioms and rules of manipulation governing the mathematical quantities are specified and when in addition certain laws are laid down connecting physical facts with the mathematical formalism.' The mathematics Dirac chooses is an algebra in which the commutative axiom of multiplication does not hold. This algebra is developed with linear operators which show this peculiar property of non-commutativeness, but which are to correspond to the ordinary dynamic variables of velocity, momentum, etc. Dirac does not make any serious attempt to give a physical meaning to his algebraic moves except of course in so far as there are variables which are interpreted as observables. For him there is no model beyond the mathematical structure of the algebra.

This kind of rarefied theory seems a very far cry from the schematic description that served as a basis for explanation in the example from ecology at the beginning of this chapter yet if we overlook the details the purposes served by theories however they are actually achieved are much the same. We have also in a way come full-circle for in the disappearance of the physical or mechanical model we get something very like the systematized description described in Chapter 3. However, the theory employing a formal model, or if you prefer it, no model at all, is not really typical of scientific explanations. By far the greater part of theory construction is typified by a movement from a simple schematism of the facts, through an informal description of a suitable model or analogy to the formal description of this model in a deductively organized system of functional generalizations, the process of formalization stopping short of a completely abstract analogy. In fact as I have described the process of theory construction in these cases some model is necessary for the essence of this kind of theory is that it does not contain a description of the facts alone, but also a description of some more familiar analogy with the facts. Since theories of this kind are much the commoner in science it is with their extension and development that I shall mainly concern myself.

Extension of a theory can run along two different but often connected lines. The extension may be what I shall call a *formal* extension, in which the logical consequences of the statements occurring in the theory are drawn out by the ordinary techniques of deduction. Extension may also be by *informal* means, in which we extend the theory by extending the model on which the theory is based. I distinguish two main kinds of model-extension, of informal drawing out of a theory, *deployment* and *development*.

The principle of an argument by deployment is that of a double use of analogy. A model depends for its efficacy in guiding our thought on bearing a double analogy to the world. If we 'think of a gas as made up of little perfectly elastic spheres' then we suppose the behaviour of the gas to be analogous to the behaviour of a swarm of small particles in an enclosed space all moving with differing velocities. And we further suppose the behaviour of the very small perfectly elastic balls to be analogous to the behaviour of the quite large, slightly less than perfectly elastic balls with whose behaviour on the billiard table we are quite familiar. In deploying our model we add to the description of the model (that is add to the model in our imagination) an additional predicate by analogy with the corresponding real entity, situation, etc. from which the characteristics and laws of behaviour of the model are taken. If one accepts the model of 'something travelling' for optical phenomena it is a natural step, by what I have called deployment, to the asking of Romer's question 'How fast is it travelling?' or that of Sir Oliver Lodge 'What is travelling?' The history of the vicissitudes of Lodge's question illustrates how hazardous is deployment for we do not know the extent of any analogy *a priori*, and so the power of deployment of any model can only be determined in practice. I said above that deployment depended upon a double analogy, and it is in the analogy between the model and the scientific information that it models that is where the analogy is wholly unknown in extent. Deployment can be represented schematically as follows:

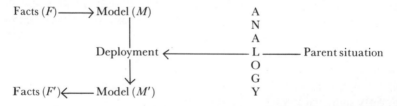

where M' differs from M in including more features of the parent situation. If M represents say the vague notion of a propagated something upon which the first attempts at a theory of optics were based, we can reach M' by adding to M the notion of whatever it is that is moving having a definite velocity, a definite mass and so on.

Development of a theory on the other hand involves the superimposing of one model on another. An example of this method of extension can also be found in optics, for the rectilinear propagation model was developed by superimposing the model of a wave without completely replacing the older model. It persists in the principle that a sufficiently narrow wave-front behaves as if it were propagated rectilinearly. Schematically represented development runs as follows:

Facts (*A*) ⟷ matched with Model I ⟵ Analogy I suggested by Facts *A*

Development

Facts (*A* + *B*) ⟷ matched with Model (I and II) ⟵ Analogy II suggested by Facts *B*

Deployment

Facts (*A* + *B* + *C*) ⟵

The further extension of the theory by the deployment of the Model (I and II) represented in the diagram above is characterized by a question like 'What is the connection between velocity of propagation and the wave-length of light?' There are other possibilities of extension too: that Model I is deployed, but in terms of the new picture we have of optical phenomena, leading to questions like 'How fast does a wave-front travel?'; that Model II may be deployed independently leading to questions like 'Are the waves transverse or linear?'; that Model (I and II) may be developed by a new analogy forcing itself on the attention of the theorist.

It must not be supposed that either deployment or development are arbitrary processes. Successful deployments of a model are not made by adding *any* feature from the parent situation (nobody has yet asked, nor could they ask, what colour are electrons). Nor are successful developments made by superimposing any model that comes to mind. Extensions are guided both from the conceptual side by certain requirements of convenience, etc. and from the factual side by their constant confrontation with experiment. Detailed discussion of these topics belongs to the second part of this book. At this point I wish simply to emphasize the fact that there is nothing arbitrary about the building up of models whose description serves as an explanation.

Once a model has become generally accepted, in the first instance perhaps for no deeper reason than through its success as a prediction device, there is a strong tendency to believe that it represents what is actually the case. Since models are used in those cases where the machinery of nature is obscure no immediate confirmation of the belief is possible but many ingenious experiments have been devised from time to time to test, albeit indirectly, the reliability and the reality of certain models. If the testing gives strong confirmation to the belief in the reality of the model, then it also gives strong confirmation to the belief that the descriptions of the model can be regarded as descriptions of the fine structure of the world. The result of a successful piece of testing of this kind is that what was an explanation of one set of phenomena is turned into something else, a description of another set of phenomena, those underlying (that is among which the antecedent conditions for the first set must be sought) the phenomena which were in need of explanation. Once this has happened it is natural to require an explanation of the transformed description of the old model. This process of progressively discovering a finer and finer structure in nature by the use of models seems to have reached an end in sub-atomic physics, for it seems that whatever model is developed its properties will derive from already understood natural phenomena. The further progress of physics may require the introduction of concepts independent of any existing system. How such concepts could be intelligible to us I believe we have no idea.

The transformations of explanations into descriptions have an important application in what I shall call *interscientific-explanation*. The explanations that have so far been discussed in this chapter were each within a particular science, that is the explanation was given within the conceptual framework in which the facts requiring explanation were stated. The structure of fish was explained in terms of the ecology of the sea, the physical properties of atoms in terms of the mechanics of their structural elements. But many explanations given by scientists are inter-scientific in that they involve reference to the facts of another science. The reference to another science takes the place of a description of a model for the description which science A gives of the antecedent conditions of certain phenomena which need to be explained is taken as an explanation of the phenomena when they are described in terms of science B. In fact the sciences can be arranged in a hierarchy such that an explanation of facts in one is given in terms of the description of facts in another, the two sets of facts being related in such a way that antecedent conditions for particular happenings in one realm are found in the other. For example certain overt pieces of behaviour in animals can be explained in terms of motives and drives, what might be called a psychological explanation, but motives and drives can themselves be given an explanation in terms of physiology, the facts of which are given a biochemical explanation, and biochemistry in common with chemistry generally is thought to be susceptible of explanation in terms of the physics of the atoms and molecules which are the units of chemistry. This is not to say that any particular piece of behaviour can be given an explanation in terms of the physics of atoms, but that the collection of more and more facts about behaviour will not provide an explanation of the piece of behaviour in question, the mechanism of behaviour must be sought out. This feature of the sciences has sometimes led to a false emphasis being placed by philosophers on the redescription of phenomena in terms belonging to another science, as if this were the central logical feature of scientific explanation[9]. It should be clear that the fact that redescription does serve an explanatory purpose is just because it has certain features in common with the intrascientific process of explanation which we have seen in central cases to depend on the construction of models. That redescription is parasitic upon model-building follows from the fact that explanations are sometimes given within a science without turning to another for a suitable vocabulary for redescription. Indeed it is arguable whether redescription with its implication of a new description of the same phenomena is a suitable expression for characterizing this kind of explanation, for it is surely characteristic of explanations that they adduce additional phenomena, real or imaginary, which are set out as the conditions of what is to be explained.

I have introduced theory with its attendant models, analogies and schematizations as if the sole purpose of this great elaboration of scientific discourse beyond immediate description was to meet our desire for satisfying explanations. However, the machinery of explanation strengthens simple descriptions by also fulfilling the general functions of codification and prediction in a more comprehensive and powerful way than straightforward description can do.

The elaboration of theory by the formal method, that is by drawing out the logical consequences of statements in it, especially when this is facilitated by mathematical expression, is more powerful than the elaboration of descriptions however effective the methods of deduction, for a theory contains more than the description of the macroscopic facts. It includes an explanation of those facts drawing on a model or in some cases the known fine-structure of nature. Formal elaboration is then strengthened by the additional premise-material derived from the explanatory background of theory. For example deductively guaranteed conclusions can be drawn both from the facts about spectra which were explained by Bohr's atomic theory and from the 'facts' about the planetary model he elaborated. Of course the conclusions reached are logically different in the sense that while those derived from the facts are themselves facts and need no 'processing', those derived from the model must be transformed by rules such as that connecting energy differences between orbits (a concept of the model) and the wave-length of the emitted light (a fact). Bearing this difference in mind we can say that the possibilities of deductive elaboration are strengthened by the existence of the explanatory part of the theory by which it differs from description. The conclusions from both description and model are, when suitably treated, useful predictions.

The informal modes of elaboration, deployment and development, described above are possible only in a theory, for they depend upon the use of models and analogies. However, they too issue in predictions though they are not in general quantitative. The move 'If light is a wave motion then there should be optical interference phenomena' is a prediction made by a deployment of the wave model in the style of (i) above. The predictions made on the basis of Fajan's theory were developments, for to the atomic model is added the notion of deformation, disturbance and reformation of the electron orbit. The hazardousness of prediction based upon informal elaboration, illustrated by the failure of the attempts experimentally to investigate the 'luminiferous ether', is just that discussed above in considering the deployment of a model, for the extent of an analogy can be discovered only by experiment. Experiments are rarely conducted, so far as I know, to determine the boundaries of an analogy, but rather in working with the analogy certain lines of investigation prosper and others peter out. The attempt to treat heat as a fluid had some initial success, otherwise it would never have been adopted as a model, but the accumulation of negative evidence, if it can be called such, from the many false starts led to its being dropped. To say 'Heat is not a fluid' is in this case to say not only that there is no calorific fluid, but to say that the analogy between calorific and hydraulic phenomena is too restricted to be of great value. However, sometimes the analogy is retained in restricted form without there being any suggestion that it provides more than a model for certain phenomena. It is still useful for example to draw an analogy between electrical and hydraulic phenomena, for the analogy goes a good deal further than it does for heat, and provides some useful models when for example electrical circuits and hydraulic networks are compared. This is an intermediate case for as I hope to show below some models are thought to be more than helpful parallels to overt phenomena.

Mach thought of science as primarily a condensation and codification of our scattered and unsystematic knowledge of facts. We have seen in the preceding chapters, how this conception though a valuable contribution to the understanding of scientific processes is too narrow, for there is a need to understand as well as a need to describe, however neatly. Of course there is a connection between condensed, codified knowledge and understanding, for only in some condensed form can we grasp the tremendous complexity of the world. Yet I believe the basic aim is understanding, and condensed descriptions are of value in so far as they subserve this aim. Condensation will be of no use unless there is a way of regenerating the information expressed in the abridgement. An ordinary generalization is a good paradigm for the machinery of abridgement and regeneration. The expressions 'all', 'every', etc. discussed in Chapter 1, can be used to say something of a range of things that is true of each thing individually. From 'All dogs are faithful' as much particular information can be regenerated as is required provided that individual members of the range 'all dogs' can be identified. The use of 'all', as well as linguistically representing the range, serves to convey the logical point that whatever member is selected from the range the predicate is true of it. The logical steps are clear:

(i) A codification procedure is followed (called in the jargon, an *induction*).

(ii) A recognition procedure is followed, by which a case of generalization is determined (based upon a *definition*).

(iii) A regeneration procedure must be employed to once more particularize the information condensed in the generalization (called an *instantiation*).*

It is important to see that the essentials of theory, schematism, model or mathematical formalism, can also be used for the abridgement, codification, and regeneration of particularized knowledge.

The abridgement and condensation of knowledge is partly a

---

* Since most generalizations simplify the data on which they are based regeneration will rarely be exact.

psychological question, but there are logical features too which are of importance in giving an adequate account of science. Models and schematisms will satisfy the conditions set out above if it is possible to generate particulars from them. In the case of simple generalizations particulars are generated directly by using the generalization or a corresponding conditional as a rule of predictive inference. With a simplified schematism, such as one finds in the diagrams used in biology, for example of the blood vascular system of an animal, there is a close logical parallel with the simple generalization for no analogy is in question except, in the whimsical sense that a drawing on paper is only an analogy of a real system of blood-vessels. A diagram, serving as a type-description of the special features of a species can be used just as a simple generalization can, to make predictive inferences from particular to particular. Imagine that a dissection is proceeding on an animal and that blood-vessels so far identified match up well with the diagram of the vascular system of the common frog, *Rana temporaria*, then from the diagram we *know what to expect* in the rest of the animal as the dissection proceeds. The phrase italicized is the usual way of expressing informally that a predictive inference has been made, and the 'rule' of inference is the diagram.

At the time when Bohr introduced his model of the atom, described above, it could with good reason have been said that the whole of atomic physics was contained in that model, together with the general atomic and kinetic theories of matter which give the Bohr atom its setting. From the model it was theoretically possible to draw out all the known experimental generalizations, for example that 'Hydrogen has a spectrum of such and such a character (the Balmer series)'. Using this generalization and others like it as rules of prediction, from a given starting point all else could, at least in theory, have been inferred. The logical point I want to make is that though a model is properly a condensation and abridgement of facts, their generation requires an additional step, the step by which the model is linked to observables, for predictive inference must logically be by means of generalizations linking observables. This way of looking at a model is the complement of the discussion of Bohr's atom for it served as an explanation too just because these links with observables could be set up. And clearly the mathematical formalism of such a theory as Dirac's shows similar logical features, serving both to explain and to abridge knowledge just because it is linked to observables. I want to say that from one point of view it is indifferent whether we retain the facts in our minds, or retain a model and know how it is linked up with observables, for the very same facts may be generated by its help. Psychologically there is an undoubted advantage in the latter procedure, for though there are dangers (we may become so taken with the model that we confuse it with the world) the simplicity, vividness and clarity of a model may be such as to outweigh other considerations.

It may be useful to sum up what I have had to say in Part One by examining briefly the various functions which the unit of scientific discourse, the individual law of nature can perform. In a way the elements of a theory; description, model, explanation, etc. are logical abstractions from a coherent and superficially homogenous body of general statements, the laws of nature. But why do we use the expression 'law' of nature? It is based upon a metaphor. Uniformities of behaviour can be represented in law-like language either because the law describes what regularly happens, or because it prescribes what should happen. If the law is followed (if the law is obeyed by individuals) this results in the uniformity of behaviour. In certain cases of human behaviour we may not know (as Wittgenstein pointed out) whether they are behaving in a certain way because they are obeying a legal restraint or because that is how things are. It might be, for a visitor from a different world, a problem to tell whether we stayed on the ground because there was a law forbidding us to rise above it. Metaphorically we can speak of objects 'obeying' the laws of nature when they exhibit various uniformities of behaviour, but there is not and cannot be a prison for those which 'disobey', for our laws do not prescribe but describe the course of nature. The term 'law' is a survival in this use of a certain theory about nature, in which there was a law-giver. Since I do not hold this theory I have used the expression 'law of nature' as little as possible. Nevertheless it is an expression in current use and does serve as a blanket term distinguishing the general statements in scientific discourse from the particulars which record individual pieces of information. Unfortunately this blanket conceals some very real differences in kinds of statement. I shall, though, continue to use 'laws of nature' as a general term for a general statement forming part of a scientific discourse.

A science enables us to satisfy three main needs, the

condensation of our knowledge, the prediction of the future course of nature, and the explanation of natural phenomena. I have tried to make it clear in the preceding pages how scientific knowledge fulfills these needs. But they are also fulfilled by single laws of nature. A law of nature, by virtue of its general form is a codification and an abridgement of material information, and our knowing the generalization relieves us of the necessity of retaining the multitudinous facts which it condenses provided that we can skillfully regenerate these facts from it. It is the general form too which makes prediction possible. As we saw in Chapter 1 any generalization is the parent of many conditionals, each of which can serve as a rule enabling predictive inferences to be made. For example 'A solution has a freezing point lower than the pure liquid' is sufficiently general to be called a law of nature. The following conditionals are among its logical progeny:

(i) If this is a solution then it will have a freezing point lower than the pure liquid.

(ii) If this were a solution then it would have a freezing point lower than the pure liquid.

(iii) If this had been a solution then it would have had a freezing point lower than the pure liquid.

I shall not elaborate the point for the situation in which each would be an appropriate rule of inference is easy enough to imagine.* Inferences from antecedent to consequent via these rules enable pre- and retro-dictions of various kinds to be made. The use of a function/variable form of a law of nature as an inference-rule was discussed above (Chapter 3). There is no room in this form of expression for subjunctive moods and perfect tenses, so that the complexity of modes of inference that can be derived from the laws of nature in simple generalization form is absent.

The function of providing an explanation is also served by the laws of nature. It will be remembered that the simplest form of explanation investigated in Chapter 1 required at least one generalization either explicitly or implicitly as a guarantee of the relevancy of one particular adduced as cause of another. In other words that one happening was the cause of another could not be claimed without the support of a generalization stating the invariable concomitance between events of the kind of the alleged

* Reference should be made here to the general discussion of Chapter 1 (p. 16 ff.).

cause with events of the kind of the alleged effect. Now just because many laws of nature are abridgements of ranges of particular facts they are generalizations capable of underwriting causal explanations. The one generalization then can perform the three functions of condensing our knowledge, justifying our predictions through its guarantee of derived conditionals which provide the rules of predictive inference, and underwriting causal explanations.

Scientific discourse is more ambitious than this, however, hoping to provide explanations not only of particular happenings but of kinds of happenings. And this involves either really giving a description of the fine structure of nature the outcome of which is the kinds of events needing explanation or failing access to this fine structure, inventing a model the description of which does the same job. Now many so-called laws of nature are not in fact descriptions of nature but of our chosen models. These can be logically of two kinds, either wholly within the model, as a statement would be which was used to describe some aspect of say Bohr's atomic model; or transformation statements, linking model and actual phenomena, as the statement of connection between energy levels and wave-length of emitted light (something more or less directly measurable) served to transform features of the model into predictions of experimentally observables. These matters have been discussed above in detail, but it is important to bear in mind that instances of all three kinds of generalizations used in explanation, the purely factual, the purely analogical (descriptive of a model), and the connective, linking model and observable fact, have been called laws of nature.* Sometimes a distinction is made between the statements which introduce a model and those which elaborate it, the former sometimes being distinguished as principles and the latter as laws, but there is no uniformity of usage in scientific language on this point. For example 'Heat is a form of motion' would, logically speaking be distinguished as a principle and 'Temperature is proportional to the average kinetic energy of the molecules' as a law. In fact this particular statement is not called a principle in the literature, the term however being used only for some genuine model-introducing statements, as in the 'Principle of the Rectilinear Propagation of Light', and the 'Principle of Least

---

* R. E. Peierls' admirable review of physics and chemistry, which he calls *The Laws of Nature*, contains as might be expected statements of all three kinds.

Action'. Though the current vocabularly is unsystematic the logical differences are I hope now clear.

Since this chapter was written (1956/7) a long running controversy has broken out over the nature of scientific explanation. It turned on a persistent attempt by logicist minded philosophers, influenced by logical positivism, to defend an account of explanation based wholly upon a deductive prescription of the form of an explanatory discourse. It now seems clear that the mere fulfilling of a formal prescription does not serve to pick out explanations from other discourse-types. As argued in this chapter, explanations are of various logical forms, some deductive, some analogical. It is the content that makes a discourse explanatory, that is it must describe the processes by which what is explained came about. As I have shown, in many cases, a scientist can only form an idea of the generative process by reasoning by analogy from descriptions of processes with which he is already acquainted. Bromberger[10] and Scriven[11] highlighted another weakness in the deductive theory of explanation when they showed how explaining is a social practice concerned with the answering of questions and the stilling of doubt.

REFERENCES

1. W. B. Marshall, *Aspects of Deep Sea Biology*, pp. 334, 335.
2. N. R. Campbell, *Physics: The Elements*, pp. 120 ff.
3. J. J. C. Smart, 'Theory Construction', *Logic and Language II*, ed. A. G. N. Flew, pp. 222 ff.
4. G. Glasstone, *Recent Advances in Physical Chemistry*, p. 18.
5. After H. E. White, *Classical and Modern Physics*, pp. 517 ff.
6. S. Toulmin, *The Philosophy of Science*, Chaps. II, III, especially Sect. 2.3, p. 28.
7. P. Duhem, *The Aim and Structure of Physical Theory*, Part I.
8. J. Jeans, *Physics and Philosophy*.
9. B. Ellis, 'On the Relation of Explanation to Description', *Mind*, N.S., **LXV**, No. 260.
10. S. Bromberger, 'Why-questions' in R. G. Colodny (ed.), *Mind and Cosmos*, Univ. of Pittsburgh Press, Pittsburgh, 1966.
11. M. Scriven in *Minnesota Studies in the Philosophy of Science*, Vol. III, Univ. of Minnesota Press, Minneapolis, 1962.

Note: An enormous literature on models and their use in science has grown up since this chapter was written. Important works are:
M. B. Hesse, *Models and Analogies in Science*, Sheed and Ward, London, 1963.
M. Black, *Models and Metaphors*, Cornell University Press, Ithaca, 1962.
M. Bunge, *Method, Model and Matter*, Reidel, Dordrecht, 1973.

Part Two
Discovery and Confirmation

5 Inductive Reasoning

The analysis of the structure of the sciences in Part One will have shown how many factors go to making up a science, how many kinds of statements there are in a scientific discourse. The important differences between a scientific and any other kind of discourse are (a) the statements in the discourse are systematically arranged and (b they are not made idly but are grounded in and tested against the facts. Now that we have seen how the sciences are constructed we are in a better position to investigate the many ways in which we determine the ultimate acceptability and satisfactoriness of scientific descriptions and scientific theories. Acceptability and satisfactoriness are of course closely related to the purposes for which we construct scientific theories and give scientific descriptions of phenomena. These purposes can be summarized as follows:

(i) the codification and condensation of our isolated items of information about the world,

(ii) the prediction of the character of the event that will occur in a new situation,

(iii) the understanding of the phenomena of nature.

We have also seen how descriptions given in certain ways and theories constructed in certain ways accomplish these purposes. We have now to ask ourselves how we make discoveries and confirm theories and generalizations which will enable us to fulfill these tasks and what criteria are used in practice to determine the acceptability of such theories and generalizations.

Traditionally it was thought that there were two kinds of problem associated with the discovery and confirmation of generalizations and theories; logical problems and methodological problems. The former were supposed to be the preserve of logicians and the latter the preserve of practising scientists. The logical problem could be put as follows: granted that we arrive at generalizations and theories from the facts at our disposal by reasoning, what sort of reasoning is this? Furthermore since a generalization or a theory is to be used for making predictions it must be more comprehensive than the information upon which it was based. How then can we devise and justify a method of reasoning in which the conclusion (a generalization or theory) is more comprehensive than the premises? Methodology on the other hand was supposed to be concerned with the detail of scientific practice, taking the solution of the logical problem for granted. We shall see that in a detailed investigation no such separation can be made for the logic of discovery and confirmation is the generalization of the methodology of the sciences. In establishing this conclusion we shall begin by considering the logical problem as if it could be discussed separately, but our discussion of it will lead us to see that the only real problems in the logic of the sciences are those which cannot be separated from scientific practice.

In making a logical study of reasoning it is convenient to have in mind some model of reasoning with which various other kinds of reasoning can be compared and the similarities and differences noticed. It is then the task of the logician to explain the differences between a certain mode of reasoning and the model that he has chosen so that he can justify the confidence we have in that mode of reasoning. The model we usually adopt, consciously or unconsciously, for comparison with all other kinds of reasoning is simple deduction; in which we can identify premises and a conclusion and can state a rule according to which the conclusion follows from the premises. The whole complex is called a deductive argument. An argument which follows a well-established rule is said to be valid; and we judge arguments acceptable or unacceptable in so far as they are valid or invalid. On the other hand we might take the notion of validity as primary and judge rules to be acceptable or unacceptable in so far as they are the rules behind valid or invalid arguments. It doesn't much matter for my purposes here which element we take to be primary, so long as we recognize the existence of these inter-related elements; argument consisting of premises and conclusion, and rule which guarantees all arguments of the same general form.

A deductive argument can be used to perform two main jobs. It can serve as a device for drawing a previously unknown conclusion; that is given the premises and the rule we can infer the conclusion. Used in this way I shall call an argument a *discovery-procedure*. When we already have a conclusion in mind, finding a deductive argument that entails it is called a *proof-procedure*. It can also be used however for determining the truth of a conclusion if we know the truth of the premises, for in valid arguments if the premises are true the conclusion must be true. Used in this way I shall call an argument a *confirmation-procedure*. The great advantage of deduction is that in the same form of argument we have both a discovery-procedure and a confirmation-procedure, so that when we draw a conclusion by means of a valid argument from true premises we know not only the form of the conclusion but also that it is true. How far do inductions, reasoning from say experimental evidence to generalizations, match up with this ideal model of reasoning?

Typically an induction, analysed on the model of deduction, has particular statements for premises, say 'Event a_1 of type X is followed by event b_1 of type Y', 'Event a_2 of type X is followed by event b_2 of type Y' and so on; and an unrestricted general statement for conclusion 'Events of type X are followed by events of type Y'. Immediately there springs up a logical problem, for according to the ideal model of reasoning, deduction, arguments of this form are invalid, so that we can neither infer this conclusion nor know it to be true on the basis of the truth of the premises. And yet, it is sometimes alleged, the reasoning in science which leads to the discovery and confirmation of the laws of nature is typically of this pattern. The logical problem, how to justify this type of reasoning, is traditionally called 'The Problem of Induction'. The reasons why this mode of reasoning is thought to produce a problem are (*a*) since the conclusion is general it will have a wider application than any collection of premises can guarantee, unless *per impossibile* infinitely many premises can be assembled; and (*b*) the truth of the conclusion can never be guaranteed from the truth of the premises to hand because a case may turn up which falsifies the conclusion. It follows then that an induction is deficient with respect to the deductive model both as a discovery-procedure and a confirmation-procedure. Since it is sometimes alleged that the demonstration of all scientific generalizations, including those picked out as the laws of nature, must follow this model we are faced with a condition of continuing uncertainty in the sciences that is at variance with the confidence that scientists feel in their conclusions. It is then thought to be a matter of urgent necessity to solve the Problem of Induction. Broadly speaking three main attempts have been made to solve the problem conceived in this way. The unsatisfactoriness of any of these attempts at a general solution will lead us to ask whether perhaps the problem has arisen through some fundamental misconception about the nature of discovery and confirmation in science; a misconception which can be cleared up and with it the appearance of an insoluble logical impasse.

ATTEMPT ONE

Logically speaking the simplest way out of the difficulty occasioned by the discrepancies between inductive reasoning and the ideal model of reasoning is to find some way to make inductive reasoning closer to the model, that is to find some device by which we can turn inductions into deductions. This can be done most easily by adding to the premises of the inductive argument a supreme premise which will be sufficiently general to overcome the difficulty of the relative width of application of the terms of the premises and conclusion and at the same time by being unreservedly true serve with the premises known to be true to guarantee the truth of the conclusion. What sort of supreme premise would fulfill the requirements for the specimen induction above? The premises in the specimen were of the form 'Event a of type X is followed by event b of type Y' and the conclusion of the form 'Events of type X are followed by events of type Y'. One possibility is the so-called 'Principle of the Uniformity of Nature', which says in effect that regularities in the sequences of types of event found in any restricted region of space and time will continue to hold good in all regions of space and time. Adding this to the premises of the specimen argument will have the desired effect for it will rule out the possibility of an aberrant case where say an event of type X is followed by an event of type Z. Hence the difficulties of width of application of terms and possible falsification of the generalization are overcome.

There are two serious objections to the adoption of this device. In the first place how can we be certain that nature is uniform and that regularities that are observed here and now will continue to be observed in other places and times? We could set about checking the supreme premise by a series of experiments but the argument which would lead from the results of these experiments to the supreme premise as conclusion would be itself an inductive argument and so another similar supreme premise would be required to guarantee its validity. There can be no end to this piling of premise on premise, so long as we attempt to prove the supreme premise by arguing from the results of experiment. Since there is such an obvious and fatal objection to deriving the premise from experiment might we not be able to prove it by argument, show it to be *a priori* necessary?

One way of deriving the premise would be to define that part of nature which is to be the proper subject of scientific investigation as the processes, phenomena, events, etc. which are uniform in space and time. Scientific Nature so defined would be the subject of an *a priori* necessary proposition 'Scientific Nature is Uniform'. It would be *a priori* in that it would not depend on whether our actual experience has been uniformly of uniformities, and it would be necessary in that its negation would be self-contradictory, given the way we have defined 'scientific nature'. But this way with the problem is too brusque. Any irregularity or non-uniformity in observations or the results of experiments would be excluded from scientific interest. But quite often the progress of science is facilitated by looking for hitherto unobserved regularities 'behind' the observed irregularities. The regularities involved in the distribution of X and Y chromosomes underlie the apparent capricious engenderings of boys and girls, who yet appear in roughly equal numbers. It seems that this way of making the uniformity of nature immune from and so independent of inductive evidence is incompatible with scientific practice. Alternatively one might regard the principle as an expression of a practical maxim, used to guide research. Its utility having been demonstrated in the past, the scientific community would be well advised to follow it in the future. The corresponding empirical principle would read 'There must be an underlying regularity even if we have not found it so far, and perhaps we never will'. In this form the failure to find regularities behind every non-uniformity would not count against the maxim, though we might have inductive reasons for holding it. It is, I think, an article of faith among scientists that all phenomena are regular and that the regularities can be expressed in laws of nature, though these may sometimes be more complex than we can conveniently state, or may be incompletely understood and so require statistical forms of law as in quantum mechanics. The existence of an apparent irregularity is taken, not as defining the limits to scientific investigation but as an indication of a certain incompleteness in the investigation of the phenomenon. To view irregularities in this way, as scientists normally do, is already to overstep the logical boundaries set by the acceptance of deduction as the model of all proper reasoning; and leads us back to the impossibility of proving by experiment this article of faith. Yet, to treat the supreme premise as a logically necessary proposition seems to be in sharp conflict with the actual practice of scientists. We are left with an insoluble dilemma; either the required supreme premise is logically necessary or it is not. If it is logically necessary it conflicts with actual scientific practice, and if it is not logically necessary it cannot be conclusively established by actual scientific practice. We can conclude then that the attempt to turn every inductive argument into a deduction cannot succeed.

ATTEMPT TWO

The first attempt at solving the logical problem broke down because of the demand that the inductive argument form should be transformed into a rigorous deductive form in which the conclusion must be true. Suppose we relax this restriction on the conclusion and instead look upon an inductive argument as leading from true premises to a probable conclusion; in such a way that the deductive model is preserved in every other respect. This requirement will be satisfied if the argument allows us both to infer the form of the conclusion and to infer its probability from the premises alone. For example if 'Event a_1 of type X is followed by event B_1 of type Y' and 'Event a_2 of type X is followed by event b_2 of type Y' and so on are taken as premises it is argued that we can, without fear of invalidity, infer as a conclusion 'Probably events of type X are followed by events of type Y'. In this form the inductive argument is too crude to be of much practical use but the form of the argument has been refined in various ways.

As a preliminary to understanding and appraising the variously refined argument forms that have been proposed we must try to get a clear idea of how the word 'probably' is used and of the various things we can mean by 'probability'. Primarily 'probably' is used in ordinary speech guardedly to make an assertion, differing in this respect from 'possibly' which is used primarily to make guarded assertions. Like the phrase 'I know' the use of 'probably' involves some kind of commitment on the part of the speaker. 'When I say "*S* is probably *P*" I commit myself guardedly (tentatively, with reservations) to the view that *S* is *P*, and (likewise guardedly) lend my authority to that view'[1]. Its use is primarily a device for indicating that the grounds for the assertion are good but less than conclusive. The job of indicating that grounds of an assertion are conclusive is done by 'I know'. It is in this sense that 'probably' is often used in the conclusion of an inductive argument, so that the hearer or reader will know that the speaker does not regard the premises as conclusive support for the conclusion. Used in this way it is clear that 'probably' does not serve to convert the induction into a deduction for it is used specifically to rule out the conclusiveness of the support which the premises may be supposed to give to the conclusion. The conclusion is in fact the same, whether we put 'probably' before it or not. In our example the conclusion is 'Events of type *X* are followed by events of type *Y*'. Guardedly to assert this conclusion is a different matter from asserting a guarded conclusion, and it is only in case the conclusion is itself guarded that the argument is transformed into a genuine deduction. 'Probably' as it is usually used serves to indicate that the argument is inductive and hence we do not remove the logical problem simply by adding 'probably' to the conclusion; rather we mark the existence of the logical problem by using this word. It follows from this that since in its ordinary sense 'probably' does not weaken the conclusion but guardedly asserts it, *i.e.* weakens the argument, senses of 'probably' which are supposed to help in removing the problem by weakening the conclusion in proportion to the degree of support provided by the premises are technical senses.

The most important technical sense of 'probability' is that which we mark by talking about the 'probability of an event'. Let us first of all explain the technical senses of 'probability of an event' and then see whether these definitions can provide us with the material for constructing arguments which are not liable to the uncertainties of ordinary generalizations from experience. There are two ways of defining the 'probability of an event', their special features determined by the method we adopt in assigning a numerical value to the chances of the event occurring.

Statistical Probability. Suppose we are investigating the composition of the forested areas in a given geographical region. We select a certain area for study and count the total number of trees in the area, say it is 678 and then count the number of oaks, say it is 337. It would then be rational to say that the chances of coming across an oak in the next area we investigate are roughly 1:2, that is we can expect every second tree to be an oak. We define a technical sense of 'probability' as follows: the probability of any given tree being an oak is equal to the quotient of the number of oaks in our sample and the total number of trees. In general the probability of any given event X (coming across an oak) occurring in a total of events Y (coming across a tree) is x/y, where x is the number of events of type X in the sample and y the number of events of type Y. Let us set out our example in the form of an argument:

The total number of trees is 678

The number of oaks is 337

$\therefore$ The probability of any given tree being an oak is $1/2$.

Here we seem to have an exemplar of deduction for the conclusion seems to follow from the premises according to a perfectly definite rule set out in the definition of statistical probability; and here too we seem to have the strength of the conclusion precisely adjusted to the degree of support provided for it by the premises.

But the appearance of deductive rigour is an illusion since there is, behind the extension of the conclusion to all forested areas in the region which alone makes the conclusion interesting, a most important assumption. The probability calculated from the results of the survey of our sample area can be applied to further areas only on the assumption that the sample area is typical of areas in the region. The illusion of deductive rigour comes from the incompleteness of the argument set out above. Properly filled out it becomes:

The total number of trees in the sample areas is 678

The number of oaks in the sample area is 337

$\therefore$ The probability of any tree in the region being an oak is $\frac{1}{2}$.

The rule of statistical inference expressed in the definition of statistical probability must be supplemented by what I shall call a *sampling assumption* to the effect that the sample area is typical of areas in the region. The only way of being absolutely sure of the correctness of the sampling assumption would be to investigate every area in the region, but to do this would make the use of probability superfluous since every tree in the region would have been investigated. In practice we have rules for sampling, rules which are intended to make as certain as possible that the sample is typical, but the rules are themselves based in the end only upon experience and upon such an assumption as the uniformity of nature whose logical status we investigated above.

Various attempts have been made to provide a rigorous justification for sampling-rules but they all in fact end in some form of the principle of the uniformity of nature; and indeed must end in this way if the probability-calculus which is based upon them is intended to be used for working out rational expectations of the chances of certain events occurring. A good example of the way the underlying assumption forces itself upon us is provided by H. Reichenbach's treatment of probability[2], though any comprehensive theory of probability will and must show the same features. The supreme principle upon which the theory is based is introduced as a rule designed to license 'the inference from a given finite initial section . . . where the relative frequencies upon which the calculations of probability depend are counted . . . to the infinite remainder of the sequence'. This rule is described by Reichenbach as a *posit*. In an earlier section of his book he defines 'posit' as follows[3]: 'A posit is a statement with which we deal as true, although the truth-value is unknown'. I think that this is just another way of introducing what we are accustomed to call an assumption or working hypothesis, which, once accepted, is the foundation for the specific sampling-assumptions which the theory is designed to prove. Within the theory of probability such a rule is an axiom and needs no proof, but its acceptance is the essential requirement of the use of the theory of probability for the calculation of the chances of something actually coming to pass. Simply to treat it as an axiom does not, as Reichenbach and other authors sometimes seem tacitly to assume, enable us to overstep the boundaries of logical rigour which the form of the inductive argument commits us to. I conclude that the use of relative frequencies to form the guarded conclusion, correctly proportioned to the strength of the evidence, which would bring the inductive argument much closer to the ideal of deduction, does not solve the problem of validation which the acceptance of deduction as the model of all argument forces upon us. It simply shifts the point at which the difficulty becomes pressing from the actual argument to the justification of the sampling-assumptions which are required to make it rigorous.

Some discussions of probability[4] centre round the problems involved in inventing a good model for probability theory. But disputes about probability-models are in a sense unreal disputes from the point of view of scientific method, for they concern the proper way of thinking about the probability calculus and are not models of what goes on in the world to which the calculus is applied. For this reason we need not consider the problem of devising a satisfactory probability model here.

Deductive Probability. When we deduce the probability of an event occurring in certain circumstances we derive the numerical value of the chances not from a statistical analysis of experimentally discovered frequencies but from the consideration of the structure of the physical system to which or within which the event in question might occur. Suppose we wished to work out the chances of a tossed penny coming down heads. We could do this simply by considering the form of the coin. Since there is nothing in the structure of the penny which suggests that one face rather than another will come up we deduce that since there are two faces the chances of any given face coming up are 1 in 2, that is the probability of say heads turning up is $\frac{1}{2}$. This looks like a rigorous form of argument for which we have been searching for the probability statement follows with deductive rigour from the description of the penny and no sampling assumptions are required. It could then be argued that if we were able to describe accurately the alternative states of any physical system the probability of its being in any given state could be deduced from the proportions of these alternatives. But if the results of these deductions are to be of any use in forming rationale expectations about the course of nature we need an assumption similar to the sampling assumptions of the statistical method. In fact we need more than one assumption; for the correct proportioning of alternatives depends upon our assuming that the alternatives we have considered are exhaustive of the possible states of the system. But this assumption depends upon another, the same assumption that underlay both statistical probability and the inductive argument, namely that Nature is Uniform. Anything that is logically possible is, in the absence of further information an alternative, and the ruling out of the manifold logical possibilities in favour of a small set of reasonable alternatives is based upon a belief that the unreasonable alternatives which are logically based are in conflict with our previous experience of the uniform con-commitances of events. In the penny case the deduction that the probability of one or other of the faces turning up is $\frac{1}{2}$ tacitly rules out all sorts of logical possibilities more or less in conflict with commonsense—which is another way of saying that we rule out possibilities in conflict with the expectations we base upon previous experience. A more or less unlikely alternative not considered in the deduction is that the penny should stand on edge, but we know from experience that it doesn't happen often enough to be worth considering. But there are a great many other alternatives which are logically possible, for example that the penny should never come down, say because it just keeps on going when tossed, or perhaps because it gets caught in the undercarriage of a passing aeroplane. These cases present different kinds of logically possible but unreasonable alternatives; the former because it is in conflict with well-known laws of nature, the latter because though not in conflict with any known laws of nature is just extremely unlikely.

It seems from these quite elementary considerations that the attempts to strengthen the inductive argument by bringing in the notion of probability in either of its technical senses does not remove the logical problem of induction but on the contrary makes its solution yet more pressing for upon its satisfactory determination the use of these methods apparently depends.

ATTEMPT THREE

Since both solutions so far proposed seem to serve only to exacerbate the problem a radically different method of dealing with it has been proposed, notably by P. F. Strawson[5] and P. Edwards[6]. The problem arises, it is argued, from the uncritical acceptance of deduction as the model for all reasoning. Once we can break ourselves of the habit of taking deductive inference as the standard of justification for the conclusion of every form of argument we are free to consider other kinds of reasoning on their own merits. In each mode of reasoning there are standards of justification, but these standards differ. What is an appropriate standard for one form or mode of reasoning is not appropriate for other forms. I have argued in another place[7] that we force the deductive model upon ourselves by using the standard vocabulary of deductive logic to describe inductions. Inductive argument, it is claimed by Strawson and Edwards is *sui generis* with its own standard of correctness; and arguments in this form simply do not need any other justification, any more than deductions need further justification.

Strawson illuminates the point with an analogy. Suppose one were to ask oneself whether a certain action was legal. This could be determined by finding out whether or not the action conflicted with a law. It would even make sense to ask whether some proposed law was legal, for it could be tested against the Constitution. But to ask whether the Constitution, the Law, is legal is to ask a question that only seems to be meaningful since there is nothing against which it could be decided. In other words since 'being in accordance with the Constitution' is what we mean by 'legal', 'legal' when applied in judgement to the Constitution has no meaning. We can ask whether a certain generalization arrived at by induction is justified, that is whether it has adequate supporting evidence, but to ask whether induction itself is justified, that is whether the method of seeking adequate supporting evidence has adequate supporting evidence is to ask a meaningless question. When we substitute for 'induction', 'the method of argument by providing adequate supporting evidence' we get for our question 'Is the method of argument by providing adequate supporting evidence adequately supported by evidence?' This has a similar air of pointlessness to the question about the constitution. But is this method of ridding ourselves of the problem really adequate?

It would be adequate if the evidence used in inductions were evidence of the same character as the evidence we might use about induction. At first sight it would seem as if the evidence must be of the same character in the two arguments for as we saw in considering the first two attempts at solving the problem, justification of induction hinges on the testing of our alleged presuppositions like 'Nature is Uniform' against the character of the world. But we shall see that the belief in the identity of the two bodies of evidence is based upon an unconscious return to the very fallacy this attempt is designed to surmount, that is the assimilation of all modes of reasoning to the model of deductive inference.

The method adopted by Edwards and Strawson is based upon what has come to be called a paradigm case argument. We cannot question the adequacy of induction as the appropriate method of establishing general conclusions about nature since induction is itself the method of determining adequacy—it is the paradigm case of an adequate argument in certain fields, particularly in the discovery and confirmation of generalizations. We simply point to a case and say 'This is a satisfactory inductive argument for here the evidence adequately supports the conclusion' and to question its satisfactoriness is to overlook the fact that it is by comparison with arguments of this form that inductive reasoning is judged.

However, paradigm case arguments are always open to a fatal objection. They can certainly be used to convince us that people do call certain argument forms adequate to their conclusion, but what this form of argument cannot convince us of is whether people ought to accept such arguments as adequate. If someone denies that a certain vocable is a name we can easily convince him that it is by reference to recorded christenings, but it is always open to him to go on to claim that it ought not to be a name, say because it has an absurd sound or because it is easily transformed into a damaging nickname. This move puts the argument on to an altogether different plane. The claim 'Induction is not a satisfactory method of argument' can be refuted on the first level by pointing to inductive arguments that are generally regarded as satisfactory. But this does not rule out the further claim 'They ought not to be considered satisfactory'; say because they are very different from deductive arguments, or because they make unjustifiable claims about the future. The idea that the evidence for the generalization produced as a conclusion of an inductive argument and the evidence for the satisfactoriness of the inductive method are of the same general character, upon which belief the method of dissolution of the problem here proposed is based, conflicts with the obvious differences in kind between the questions framed on each level. While we consult one set of facts to find out if our generalization is justified we consult a different set of facts to find out if the method of consulting facts of the first kind is a sound method. To suppose that there is an identity of factual and methodological evidence is once again to think of inductions as following the deductive model; having premises, a rule of inference and a conclusion; for if we are looking for a supreme general premise to validate all inductions at once this must itself be founded upon factual evidence. But the proper method of justifying inductive methods is to ask how far they fulfil the requirements set by the tasks which a science is supposed to perform, and how far they fit in with the requirements of theory by leading to the acceptance of generalizations which are useful and satisfactory to us.

There is however one way of putting the 'problem' of induction to which the arguments of Strawson and Edwards do apply. If we set the problem in a completely general way; that is by asking what is the justification for consulting the facts in order to justify or discover or confirm a generalization, then Strawson's and Edward's argument serves to remind us that there is no possible answer to this way of framing the question. What other method of justification could we find? But this method of framing the 'problem' is so general as to be quite unilluminating for it cloaks the real questions of scientific method; namely what in fact are our methods of hitting on new generalizations and theories and what are our criteria for judging these generalizations and theories and why do we find these methods and criteria useful? We must give up the idea of justifying scientific method in a single bold stroke and content ourselves with a piecemeal appraisal of the logical stucture of the various methods scientists actually use. The logic of science seems to consist then in a study, from a logical point of view, of the methods and criteria actually in use.

The whole idea of there being a separate and special province for logicians in the study of the sciences is based upon some fundamental misconceptions. The most deep-seated which promoted Attempts One and Two above is the idea that inductions must follow the model of perfect reasoning; logician's deduction. Connected with this misconception is the idea that there is a standard inductive argument form, and that all we need to do is to identify this form and argue for its general validity. These related misconceptions lead to another, more superficial misapprehension; that the same standards of acceptance apply to generalizations as apply to their supporting evidence; that 'true' is used in the same sense (based upon the same criteria) for both generalizations and the particular statements which are our evidence for generalizations. We can go some way towards removing these misapprehensions by considering a pair of specimen cases in which evidence is presented for a conclusion more general than the facts upon which it is based.

A very common way of describing one kind of inductive step is to say that such and such results (or observations, information, etc.) suggests an appropriate general statement. Now it can easily be shown that when the evidence suggests a certain generalization this is not assimilable to rule-bound inference, implication, which is the paradigm of arguments of the type of deductions. Consider the various ways in which suggestion and implication can be disputed. A simple non-inductive example can perhaps make the logical watershed between the two modes of inference clear. For example we can distinguish a suggestion from an implication most readily by considering the way we would dispute their conclusions. 'The inclusion of Miller suggests that the Australian bowling is weak' can be disputed by something like 'It doesn't suggest that at all, it suggests they need more batting strength in the middle of the batting order' for it is the adduced facts which suggest the generalization, there is no third element to be disputed. Provided the facts are accepted the dispute concerns the conclusion directly. Argument about what the facts suggest reduces in the end to attempts by each party to get the other to 'see' the facts in a certain way. It is significant that we nearly always ask 'What does so and so suggest to you?' Argument about a 'law of suggestion' would be superfluous in this dispute for if suggestion depends directly upon how the facts are seen, then disputes about the result would be able to be carried on independently of the 'law', so that it would be idle. But 'From the inclusion of Miller we may infer that the Australian bowling is weak' can be disputed by arguing against the principle of the inference, which would be a dispute about the rules of selection for cricket teams. For example it could be said in dispute of the above inference 'Weakness in bowling doesn't necessarily mean that you include all-rounders' and this would be a move against the inferred conclusion indirectly, but directly against the principle according to which the conclusion was drawn.

The facts suggesting a generalization is evidently fundamentally different from the generalization being deduced from or even inferred from the facts. Inductive procedures in fact form a spectrum ranging from informal intuitive steps such as the facts suggesting a generalization or theory to formal devices such as the methods of probability and functional generalization which are closely similar to deductions in form. The idea that arguments of deductive rigour must be our model for all forms of reasoning springs I believe from a concentration upon inductive methods at the formal end of this spectrum; what I shall call *mechanical* inductive procedures; to the exclusion from our consideration of the many cases in which we would describe our mode of reasoning by the use of some such word as 'suggests'; modes of reasoning which I shall call *intuitive* inductive procedures. When it comes to the question of the acceptance or rejection of an hypothesis or hypothetical theory the criteria that are in fact used form a vast family of many different kinds. There is an apocryphal story about Einstein's methods which illustrates this point. He was once shown a very powerful physical theory which successfully accounted for all the known facts in a certain field of investigation. To the annoyance of the author of the theory, Einstein, after glancing at a few pages, handed it back with the remark that the theory couldn't be any good as it was insufficiently beautiful. There are two main kinds of criteria used for appraising hypotheses, criteria derived from the basic requirement that our generalizations and theories 'fit the facts' and criteria derived from the quasi-aesthetic and logical properties which we have found it desirable for our generalizations and theories to possess. In the next two chapters I shall make a brief study of some of the methods of judging generalizations and theories based upon criteria derived from both these sources.

A characteristic feature of deductions is that given the truth of the premises, if the deduction is valid, the conclusion is also true. This is one of the features in which inductions differ most markedly from the deductive model, for the 'conclusion' of an inductive argument is always wider than the information contained in the 'premises'. One way of putting the old logicians' problems of induction is to see it as an attempt to bridge the gap between the truth of 'premises' of an induction and the relative uncertainty of the general 'conclusion'. But why should we judge generalizations and theories according to the same standards as we judge particular statements of restricted application? Scientists, in fact, most often use the expression 'satisfactory' to mark their approval of a generalization or theory. Satisfactoriness for them seems to be a very complicated matter, different in several important respects from the truth of simple particular statements of fact. We shall have some idea of what makes a generalization or a theory satisfactory to a scientist when we have studied the wide range of criteria by which we determine whether to accept or reject an hypothesis. If the 'conclusions' of inductions are judged by different criteria from their 'premises' there may not in fact be any gap to close and this way of putting the problem of induction another aspect of the obsession with the deductive model.

From time to time in this chapter I have made use of a distinction among inductive procedures between methods of discovery and methods of confirmation and rejection. In deductive arguments these aspects of the argument are not distinct for when we draw a conclusion from true premises according to a valid inference form we know the truth of the conclusion too. But in inductions we make hypotheses and then we test them, our discovery-procedures and our testing-procedures frequently being distinct activities. An hypothesis suggested by a meagre accumulation of facts *may* be tested by the systematic piling up of instances as for example in the testing of drugs and antiseptics. Logicians in the past were, more often than they were perhaps warranted, interested in those methods by which the hypotheses are derived from the collection of particular information and their testing carried out by the further accumulation of instances, the method Bacon approved. A strong corrective to this narrow view has come from the ideas of K. Popper[8]. He argues that scientists are often not concerned with whether a hypothesis or hypothetical theory should be proved to be acceptable but with the possibilities of its rejection. His view could be summarized in this way: we begin with certain vaguely formulated ideas about the way things are and how the world works, giving us a certain horizon of expectations about what will happen in given circumstances. This we make more precise by constant refinement as we find reasons for rejecting those parts of our vague view of the world which do not fit in with our further experience. Another way of putting this is to say that laws of nature are always provisional, their temporary acceptance meaning that we have not yet found any grounds for their rejection. This reversal of our customary direction of viewing scientific method does not, as Popper once thought, solve the problem of induction, but does serve to draw our attention to a very important fact, that in many cases discovery of new hypotheses is not by induction at all, but by the development of theories we already possess. The growth of science is a complicated matter, new laws of nature being discovered both by extending and refining our theories and by generalizing new factual information. We judge some of our hypotheses satisfactory both because we find plenty of confirmation of them and because we do not come across cases that would call for their rejection[9]. In the next two chapters we shall investigate in detail the logical structure of some of the methods by which the sciences grow and some of the criteria by which generalizations and theories are tested.

REFERENCES

1. S. Toulmin, 'Probability', *Essays in Conceptual Analysis,* ed. A. G. N. Flew, p. 165.
2. H. Reichenbach, *The Theory of Probability,* §§87, 91.
3. *Ibid.*
4. R. B. Braithwaite, *Scientific Explanation.*
5. P. F. Strawson, *Introduction to Logical Theory,* p. 257.
6. P. Edwards, 'Bertrand Russell's Doubts about Induction', *Logic and Language; Series I,* ed. A. G. N. Flew, Chap. IV, p. 55.
7. R. Harré, 'Dissolving the Problem of Induction', *Philosophy,* **XXXII**, No. 120.
8. K. R. Popper, 'The Philosophy of Science', *British Philosophy in Mid-Century,* ed. C. Mace, p. 155 ff.
9. Since this chapter was written great efforts have been put into developing Popper's 'fallibilism'—see particularly I. Lakatos and A. Musgrave, *Criticism and the Growth of Knowledge,* Cambridge University Press, Cambridge, 1970.

6 'Fitting the Facts'

In giving an account of description in Part One I gave at the same time of necessity an account of the formation of descriptive generalizations. I should like at this point to turn again to the procedures used for forming descriptive generalizations in order to make the account of discovery and test procedures complete. In the last chapter I distinguished between mechanical and intuitive discovery and testing procedures in a general way. This distinction can be given a detailed application in the business of giving in the most scientifically desirable form descriptions that 'fit the facts' and of testing these general descriptions to discover their correctness and the limits to their field of application.

When our chosen observables are quantifiable the results of experiment and observation can be expressed in tables in which the numerical values of the observables under different conditions can be recorded. It will be remembered that the analogy between an algebraic variable and a changing quantified observable allows these tables to be condensed into suitable algebraic functions thus giving in an extremely economical form a general description of the process in which the observable plays a part. A characteristic example of an induction procedure which leads to a functional generalization from numerically expressed particulars is that of the generalization which is produced by dropping from an algebraic function representing the known facts, terms of higher powers. The very idea of 'the known facts', upon which the analysis is based will be scrutinized in Chapter 8.

In order to construct an example we can consider the process of the formation of a functional generalization as made up of two sub-processes. As a first move we simply aim to record the information we have without extrapolation, that is without going beyond this information. Algebraically this can be done as follows: a table of results, say of the measurement of voltage and current in a conductor, the example of Chapter 3, might be:

V	2	3	4	5	6	7	8
I	3·93	6·01	7·97	9·80	12·11	14·03	15·98

We shall imagine the voltage to be variable experimentally and the corresponding current to be read off an ammeter. In Chapter 3 we imagined that there were no experimental deviations as a comparison with the table on page 56 will show. This table is however nearer to the kinds of results we would obtain in practice. The results in the table can be exactly represented in an algebraic function by forming the function from seven factors (the number of pairs of observations) each factor formed from a pair of observations so that it will represent a solution of the algebraic equation formed by the product of the factors equated to zero, *i.e.*

$$(3·93V - 2I)\ (6·01V - 3I)\ (7·97V - 4I)\ (9·80V - 5I)\ (12·11V - 6I)\ (14·03V - 7I)\ (15·98 - 8I) = 0 \qquad (1)$$

This equation is of the form

$$aV^7 - bV^6I + cV^5I^2 - dV^4I^3 + eV^3I^4 - fV^2I^5 + gVI^6 - hI^7 = 0 \quad (2)$$

and giving the coefficients $a, b \ldots h$ their numerical values from (1) the seven stationary values of this equation reproduce the seven results from which it was derived.

A functional generalization of this kind has several undesirable features which make it of little use as a scientific description of the flow of current in a conductor.

(i) Though in the form of a functional generalization it is not a true generalization. Any additional information will require the addition of an extra factor for each pair of results so that the function will become algebraically of progressively higher power as the amount of information increases. As the algebraic process by which it is formed leads only to this pseudo-generalization I shall call it a pseudo-induction.

(ii) It is clearly very unlikely that a function of this form, even if it is restricted to a definite number of factors, will be derivable from others of different form and no doubt of a different number of factors and so of different power. One might say that the algebraic form is arbitrary, for it depends upon another arbitrary factor, namely how many readings of the observables one had made. This feature automatically frustrates that fundamental desire for system that was noticed in Chapters 3 and 4 and which conditions so much of scientific method.

A direct method of induction from the results of a similar imaginary experiment was described in Chapter 3 (page 47), but without considering how the complexities that result from 'untidy' experimental 'data' are dealt with. The polynomial (1) above differs from $V = IR$ in that the measured values of the dependent variable I differ in an apparently unsystematic way from the corresponding values of I given by the idealized law. By finding the sum of the squares of the differences between the measured and the ideal values for I at each value of V a statistical measure of the plausibility of choosing the idealized form can be calculated. A similar account can be given of the drawing of smooth curves as the best 'fit' to a plot of actual experimental results. These are true inductions.

Sometimes 'untidy' experimental results can be systematized by including higher powers of the dependent variable. The tentative form of a law might appear as

$$V = IR + I^2S + I^3T + \dots \tag{3}$$

This move presents theoreticians with the challenge of finding physically plausible interpretations of S, T, etc. The effect of the addition of the higher power term is small. Unless a suitable and compelling interpretation is available these terms are often neglected. I owe to Dr I. Aitchison the important observation that in some branches of physics the radical induction of polynomials that I have described here is discarded because the 'interesting' results are to be looked for in the higher degree polynomials that come very close to the purely 'phenomenological' descriptions of pseudo-generalizations like (1) above.

Testing by the facts is the complement of discovery, of using the facts to derive a generalization, for any discovery-procedure makes an advance upon the facts already known in that while the facts are of necessity particular the procedure issues in descriptions which are intended to be general. Just because they are general they are hypotheses, for they are extrapolations beyond what is already known. It follows then that acceptance of a hypothesis reached by any discovery procedure which depends upon generalizing must depend heavily on the confirmation of predictions made by means of it. It is a commonplace that no generalization can be tested directly but only through the success or failure of some particular prediction. This must involve a deliberate experiment. In using a generalization as a rule of inference in the conditional form, or if it is algebraically expressed through the substitution-calculation procedure, to make predictions premises are required from which according to the rule the prediction is made. These premises, stating particular facts, also set the experimental situation in which the prediction will be tested, for a prediction always takes the form: In the circumstances so and so (described in the premises of the predictive inference) such and such will occur (described in the conclusion of the predictive inference). To make the testing of the prediction possible nature must be manipulated in such a way that the circumstances described in the premises are brought about, when the appearance of the circumstance described in the conclusion confirms the prediction, and its failure to appear disconfirms it.

A schematic representation of the whole process of mechanical induction would run as follows:

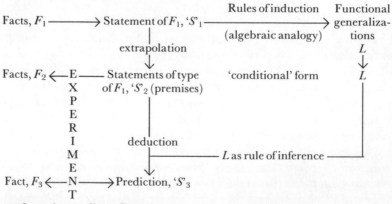

A prediction can easily be made with the true generalization obtained above. If we suppose that say, $V = 9$, then using the generalization as a rule of inference in the manner described in Chapter 3 (page 67) we infer that $I = 18$. An experiment can now be devised. By the necessary manipulation of external resistances we make the voltage in the conductor 9 volts, and then read off the value of I, which we expect to be more or less near 18 amps. Coincidence between inferred value and experimentally produced value, within the range of error set by the formation procedure of the generalization used as an inference rule, constitutes experimental confirmation. The lack of coincidence, outside the limits of error, constitutes disconfirmation. This simple scheme is found very widely in science, and is perhaps the basic scientific method.

The discovery and testing procedures I have been describing are typical of most respectable scientific investigations that are undertaken nowadays but it is still important to bear in mind that these methods are not radical departures from the ways we ordinarily generalize and the ways we ordinarily 'make sure'. They are, rather, rationalizations of procedures with which we are quite familiar. A clear case with which everyone has some acquaintance is that of the drawing of graphs. The graph is, or represents, a generalization of the information to hand, for the graph is continuous (an unbroken line) while the information is finite, limited and discontinuous (dots or crosses on the paper). Graph drawing can now be made fully automatic, for machines now exist which express in graphical form the trend of information fed into them. This is indeed the paradigm of mechanical induction. But it is remarkable how closely this follows one of our ordinary intuitive 'methods', for spotting a trend can be greatly facilitated by representing the information we possess in some visual way, say by dots on paper, and letting the eye make its own survey. The *gestalt* of the plotted points, the pattern in which we see them, is, more often than not, the best expression of their trend, making a pattern which would be reproduced by the operation of a computer. Similarly we can look upon probability-theory and statistical-analysis as elaborate ways of making more precise and quantitative the generalizations and judgements of generalizations with which we fill our ordinary conversation about the world. Numerical probabilities correspond to the 'more or less likelies' of qualitative appraisals and statistical devices such as mean values correspond quantitatively to those vague expressions in which we ordinarily express generalized information, *e.g.* 'the average so and so', 'most people', 'the man in the street', 'anyone will tell you that . . .' (which has been quantified by the consumer research firms), and so on. Elaboration of these devices for making precise generalizations would be out of place here, for I do not wish to concern myself with the detail of scientific methodology. They serve however to illustrate the point in logic I wish to make.

Many generalizations that later come to undergo detailed testing are first derived from particular information, not in one of the ways outlined above, which are developments of our usual intuitive procedures, but rather in one of the many ways in which in quite ordinary circumstances we reach the general from the particular, those ways which are rationalized in the mechanical procedures discussed above. It is often of great practical help to make some rough estimate of a constant, some rough guess at a correlation between factors, which, though the support given by the evidence on which the estimates and guesses are based is exceedingly difficult to formulate, are not independent of this evidence. For example it is quite common to find in the scientific journals phrases that suggest a certain correlation between variables was suspected or guessed at long before it could be said just what the correlation was. There are many examples in the history of science. For perhaps a hundred years it was clear that force, mass, velocity, etc. were closely connected, but just what the connections were was only gradually established by the work of many men and through the difficulties of many false starts. Of course there is much less blundering in the dark now, for there is the powerful tool of the calculus of correlation coefficients to help us; even so this tool can be effective only when our information has already been expressed quantitatively, that is it can only be used when the correlations established are between observables that are either countable or measurable.

In all these cases, whether the generalization is hit upon by a *coup d'oeil* or by mechanical means, or in any other way, it is arrived at by the turning of our attention away from particular detail to a scanning of the material on the look out for something in common. Certain details, those characteristic of individual instances, if it is a class of objects we are scanning, must be passed over unconsidered if an inductive move is going to be possible. The first example of this chapter shows this point very clearly for the generalization we might say was created by ignoring the small experimental deviations that were characteristic of the results of individual experiments.

There must though be some criteria for deciding when a deviation from strict regularity is small enough to be passed over and when it is sufficiently great to be of importance. The decision will depend upon two factors; the first how reliable one regards the apparatus used, and what in fact its margin of error is likely to be;

secondly whether the current theory in the field in which the experiments are being carried out allows for any development that can account for a certain sort of error, or irregularity. Very often these two factors combine in making the decision as to what is an important and meaningful irregularity. This may be either because the current theory suggests that there are additional factors which have not been taken into account in the earlier generalizations in this field of investigation so that deviations from these too restricted generalizations are to be expected, or because the deviations that are in fact found are so great that experimental error cannot account for them and theory must be modified to meet them.

An example of this sort of decision is the deviation from the classically calculated position of the planet Mercury in the observations which provided a confirmation of Einstein's theory of gravitation. He predicted that the precession of the orbit of this planet would differ from the precession calculated according to the then current theory by a certain quite definite amount. This amount was far larger than the experimental error which could be put down to the apparatus. If the theory had predicted a deviation less than the experimental error inherent in the apparatus of telescope, clocks, etc. then it would and could have had no immediate significance. No immediate significance, but no significance at all, for the appearance in a science of predictions that represent differences of finer grain than can be detected by the apparatus to hand, is a powerful stimulus to the design and construction of apparatus in which such differences become significant. On the other hand when Amagat used the gas law apparatus he had built in the shaft of his father's mine the deviations he found from the well-known and currently accepted laws were far greater than could be accounted for by any error in his measurements or inherent crudity in his apparatus. His experiments together with those of Andrewes on carbon dioxide led to the development of theory propounded by Van der Waals, in which the size and mass of the elementary particles of a gas, hitherto not significantly imported into the theory, were used to account for the irregularities of behaviour relative to the current laws that had been observed in these experiments. I shall have occasion to return to this point later, but for the moment it can be said that the character of our apparatus sets a threshold beyond which deviations cease to have significance.

II

The building up of a theory and the invention of an appropriate model are closely related activities to which the notions of discovery-procedures hitherto discussed in this chapter do not seem to have any immediate application. We do not think of a theory as derived from the facts, but rather as brought to bear upon the facts, as adding significance to them, as when we talk of giving an explanation. This is not to say that facts are irrelevant to theory-construction but that they do not of themselves constitute an adequate basis for it. The extra that is required is an appropriate model, description of which, linked to the description of the facts the theory is intended to cover, provides, as we saw in Chapter 1, the basis of scientific explanation.

The facts, we might say, suggest a theory, in so far as there are certain features of the relevant facts which have a parallel in another situation. The parallel provides the basis for the use of this other situation as an illuminating analogy and the development of the parallelism between facts and linked situation (the model) marks the scope of the theory. This can best be seen in an example. A well-known model, the atom modelled on planetary systems, was 'discovered' in the following way: Rutherford was investigating the passage of small particles through very thin metal foil. He found that there was little random scattering but that the characteristic features of the experiment were an unrestricted passage through the metal for the majority of particles while an occasional particle was reversed in direction so violently that this reversal could only be explained by supposing that it had collided with something very heavy. These features parallel the features that would be characteristic of bodies meeting planetary systems, the great majority passing harmlessly through the empty space in which the planets move, and the rare instance striking the nucleus itself and being reflected violently back upon its path by the force of the collision. Once Rutherford saw that the parallel existed the analogy developed rapidly with the atom being given a nucleus around which revolved the planetary electrons in their appropriate orbits.

Whatever other criteria we use for testing theories, no theory can be regarded as successful if it is in flagrant breach of the facts. I use 'flagrant' here advisedly for there are other criteria, simplicity is one of them, for the satisfaction of which we are inclined to sacrifice something in the way of correctness, of exact match with the facts. That there are good reasons for making these exceptions we will see in the next chapter. A collision with fact serious enough to effect the ultimate appraisal of a theory must satisfy two requirements:

(i) it should not be considered a disconfirmation if the difference between theory and experiment is less than can reasonably be expected with the apparatus in use;

(ii) it will be a disconfirmation only if it is not possible to account for the discrepancy by some development or deployment of theory without the making of major modifications.

This rules out the use of certain merely *ad hoc* (or arbitrary) additions to theory to prevent its disconfirmation. Not being in conflict with any known facts and being able to be used to make predictions that can be checked in fact are the negative and positive aspects respectively of theory testing. We may say that a theory is minimally successful provided that it is compatible with what is already known, and that any prediction made by means of it is successful.

A theory cannot itself be tested but must be reached indirectly through the success of the generalizations that can be derived from it. Nor can a generalization itself be brought to direct test, for the facts with which we can confront our constructions must be always particular. We must express the generalization in its rule of inference form and so use it to make a prediction of particular fact which can be directly confronted with the world. The processes of forming new generalizations from theory, either by deduction or by development or deployment have been described above (Chapter 4) while the testing of a generalization has been described in Part I of this chapter. I shall try to bring these scattered elements together in three examples.

1. As an example of a deductively related system of generalizations I considered in Chapter 3 the laws of elementary kinematics. Here we have a deductive system of symbols that is interpretable as kinematic laws. Starting with

$$v = u + at$$
$$s = ut + \tfrac{1}{2}at^2$$

by the ordinary rules of algebra, another law can be deduced.

Eliminating t we have

$$s = \frac{u(v-u)}{a} + \frac{a(v-u)^2}{2a^2}$$

$$v^2 - u^2 = 2as$$

This new generalization can be used as a rule of inference, for given, say, that $v = 4$ m./sec., $u = 3$ m./sec., and $a = 1$ m./sec./sec. it is easy to deduce that

$$s = 3.5 \text{ m.}$$

An experiment could easily be arranged to measure how far a trolley moved in being accelerated at the rate of 1 m./sec./sec. from 3 to 4 m./sec.

Since the system of laws that make up the theory is deductively related, the success of the individual laws as inference rules reflects upon the satisfactoriness of the system from which they were derived and of which they form a part. From true premises with a correct use of the rules of deduction we can deduce only true conclusions, and a false conclusion, an unsatisfactory generalization, throws doubt upon the whole system of laws. Of course the confirmation by the success of a prediction that attaches to a single generalization even if deductively related to the original system does not provide a simple guarantee of satisfactoriness, for any of the premises may be false and yet compatible with the truth of a conclusion. There are many forms of argument in logic where this is so. It is this vagueness of confirmation that is partly responsible for the use of criteria other than the factual in determining the satisfactoriness of a theory.

2. Fajan's theory used in Chapter 4 to exemplify the informally developed model can be deployed to provide new generalizations which can be used to make predictions. The model, it will be remembered, was that of the Rutherford atom supplemented by a special hypothesis concerning the effect upon the orbits of the planetary electrons of two ions being in close proximity to each other. The immediate requirement, that the theory account for the anomalous behaviour of beryllium chloride is satisfied by the bare introduction of the model. But without introducing any new hypothesis the model leads to various predictions. This is what I have called deployment. For example it may be predicted that compounds formed from ions with large unbalanced charge will tend to be covalent, for this large charge will tend mutually to

distort the electron orbits of the combining ions so that they tend to be combined by sharing rather than losing electrons. In terms of the periodic table of the elements compounds formed from ions of elements occurring in the middle of a period will tend to be poor conductors for such elements will tend to lose what electrons they have thus producing ions of high unbalanced charge. We can in fact predict that compounds formed from elements whose ions have large charge and small volume, that is which are formed from elements occurring in the middle groups of early periods will tend to be covalent. A typical covalent compound such as carbon tetrachloride is formed from carbon and chlorine, the former occurring towards the middle of Period II, and the latter towards the end of Period III. The carbon ion then has smaller volume and greater unbalanced charge than the chlorine, and so will have the effect in close proximity of distorting the orbits of the electrons of the chlorine ion to such an extent that they are drawn in revolution around both ions. The conditions for covalency are exactly fulfilled. Logically speaking the model has led to a new generalization 'Ions of small volume and high charge tend to distort the orbits of electrons of neighbouring ions', which can be transformed, by substituting for the model-descriptive expressions their counterparts in fact, into a generalization about observables such as chemical properties and electrical conductivities. For example an acceptable transformation yields 'Elements which occur in Groups III and IV of Periods II and III of the table will form compounds which are not readily soluble in water and which are poor conductors of electricity'. By means of this generalization predictions of particular fact can be made. The success of these predictions tends to confirm us in our belief that the model is a good one. But reflection shows that on these grounds 'good' can be given no more content than say 'successful as an aid to prediction'.

3. What I have called development of a model (Chapter 4) consists in adding to the basic picture an additional or supplementary model, which while consistent with the original picture transforms it into something much more powerful as an instrument of explanation, and equally of prediction. Examples are not hard to find; consider the way the original simple picture of the atom which Rutherford invented has been supplemented and supplemented again each time proving an expanded basis for the making of new predictions. The example in Chapter 4 of a

formally developed theory is also an example of a development of Rutherford's model for new but consistent features were added by Bohr. His main addition was the idea of a series of fixed orbits in which the electrons were constrained to move, their passage from one to another eluding explanation or even proper picturing at that stage of the development of the model. But this development even with its obvious deficiencies as an explanation did enable great advances to be made for generalizations about energy emission and spectrum patterns could be deduced, and the detailed predictions that were made in accordance with them were spectacularly confirmed. To say that Bohr's atomic model is an improvement on Rutherford's has at least this minimum sense, that as an instrument of prediction it is more powerful and more precise.

The power to predict and the power to explain are closely connected. In judging an explanation we do not rely entirely on its power to satisfy us that we have understood, for at least part of what we mean by understanding can be rationalized into a simple logical relationship. 'Does theory *T* explain event *x*?' can be partly expressed as 'Does theory *T* permit a mock-prediction of event *x*?' I use 'partly' here for there are other criteria and a successful theory does not just have to be a successful prediction device. The power of prediction, what might be called the logical power of the theory, is a necessary but not a sufficient condition for a favourable assessment. Corresponding to real predictions which end in putative statements of facts not known before are mock-predictions, whose purpose is to show that what we do in fact already know could have been predicted had we had the theory. And this is part of what we mean when we say that a certain thing has been explained. 'Does theory *T* explain event *x*?' can be answered affirmatively if in one of the many ways discussed in this chapter theory *T* via generalization *G* allows us to predict event *x*, provided that theory *T* is itself an acceptable theory. It must give us the feeling that we understand why event *x* occurs. 'Does the Fajan theory explain the low conductivity of beryllium chloride?' Yes—provided that we accept the theory, and that the low conductivity can be derived from the theory, that we can say that we would expect beryllium chloride to have a low conductivity. I hesitate to use the word 'infer' here to describe the relation between theory and prediction, for theory does not provide the premises of inference, but the rule. Furthermore it carries too a suggestion of a simple deductive relation that seems quite inappropriate in those cases where the prediction is the outcome of the use of a rule of inference derived either by development or deployment of a model.

These examples and methods of confirmation have all been cases of indirect testing of a theory against actual fact. The theory gives us the generalizations in one of several ways, the generalizations are transformed if necessary from the terms of the model to the terms of the appropriate observables and are used to make predictions of one matter of fact on the basis of other matters of fact. It is only the prediction that is confronted by a direct empirical test, and it is only of the prediction that it can be asked 'True or False?' An extreme positivist might argue that this is a perfectly satisfactory state of affairs provided we do not inflate our conception of a scientific explanation too highly. The model is a crutch for the stumbling intellect and can be dispensed with so that if we look upon a science as no more than a summary of those individual pieces of information we have already collected and a device for generating more information of the same sort then a 'good' theory can mean no more than a 'successful prediction device'. Such a view is salutary since it drawns our attention to the dangers of taking models too seriously, but it is for all that too narrow. It is unquestionably the case that a theory will stand little chance of acceptance if it does not have some appeal to our desire for understanding, and the satisfaction of this desire, as we saw in Chapter 1 is closely bound up with the use of a convincing and stimulating model. The building of a theory and the growth of one's model form a single process in ordinary research work, for good practical reasons, but for logical purposes I propose to treat the process in steps or stages. I shall preserve this logical fiction in the discussion that follows, though I think that in some extreme cases these stages are more than a fiction and are indeed the actual procedure adopted. I shall suppose then that a model is first invented, then tested and finally in a summing up of the results of the tests accepted or rejected.

The examples I have discussed above have turned on our using the theory as a prediction device and on our treating the model as no more than a part of this. But it is clear from the conversation and writing of scientists that this is far from their ordinary view. There seems to be a widespread supposition that the model that is invented is also, if it is acceptable on predictive and other grounds to be described below, a representation of the logical fiction above it seems that many scientists look upon their models as candidates for reality, so that we have not only to test them in the ways I have described, but also to find out how far they do hit off the finer structure of nature. Some very ingenious experiments have been designed with exactly this aim in view. Sometimes they have been successful and vindicated the model by exposing a hidden mechanism, a finer structure in nature, which is identical or closely similar to that postulated in the model. Occasionally the experiments have failed and a new model has been forced upon us, indeed a new view of the structure of things has sometimes been necessary to overcome the uneasiness created by the demonstration that some model or other is not in fact acceptable as an approximation to the real underlying structure of the world. This kind of testing I call *exposing* the model.

There are many excellent examples of experiments designed for the purpose of exposing a model. I shall take an example from recent experimental work in hydrodynamics. Until recently it was supposed that the boundary between any two phases of a physical system was characterized by the existence on either side of the boundary of a stagnant film of small but determinate depth. The belief in the existence of this film as an actual physical reality was at one time complete and universal. The film played a crucial part in the theories of transfer across phase boundaries, for in a stagnant film the mechanism of transfer of say heat, must be molecular diffusion and cannot be convection for the latter depends upon a movement among the elementary particles. Such movement is ruled out by the stagnancy of the film. Some experiments on the structure of the so-called stagnant film were carried out many years ago by Fage and Townend[1] the significance of which has only recently been appreciated by theoretical workers in this field. By means of an ultramicroscope Fage and Townend investigated the state of a liquid very near the phase boundary and found that within the limits of their experiment which were finer than the theoretically predicted depth of the stagnant film there was no laminar sublayer. There was in fact a layer just below the surface in which there was movement among the elementary particles both laterally, parallel to the phase boundary, and normal to it. There is evidently a contradiction of the theory that the sublayer is stagnant and seems to be a clear experimental refutation of the objective reality of a certain model. A theoretical worker on boundary transfer, Hanratty[2], has cited these observations among others in support of a theory of boundary transfer that makes use of convection as the major transfer mechanism.

There seem to be two conditions for this kind of theory testing to be possible. The first is that the model upon which the theory depends should be neither purely formal (*e.g.* a mathematical calculus) nor purely analogical (the water-drop model of the nucleus of atoms) but should be expressed in such a way that it could conceivably *be* real, and that the description of it which figures in the theory and which we use to assist us to grasp the processes etc. which are explained by the theory could be a description of the fine structure of reality. The water-drop model of the nucleus is ruled out by this condition for it is logically impossible for the nuclei of atoms really to be composed of water. The second condition is that there should exist instruments capable of detecting the existence and measuring or at least marking the properties of the entities postulated in the model. In the example above the ultramicroscope fulfils this second condition, for by detecting the light reflected from exceedingly fine particles suspended in the liquid being observed it enables very minute movements to be observed in a region well within the theoretically stagnant layer.

There are other procedures for determining the satisfactoriness of models that do not to the same degree commit themselves to maintaining the actual existence of the hidden mechanisms postulated in theory. I shall discuss two examples, one in which the question of the model's existential status is left open and another in which the model is specifically treated as no more than a useful device.

1. Difficulties were encountered in early investigations of radiation in giving a satisfactory account of X-rays through the lack of any definite means of distinguishing these radiations from fast-moving particles. A standard test for a wave motion is to determine whether the radiation can be diffracted. It was realized that if indeed X-rays were waves they must be of such short wavelength that no known grating would be sufficiently fine to diffract them. Von Laue made the suggestion that crystals, if they were in fact lattices of regularly spaced molecules, would serve as diffraction gratings of the requisite fineness. Using crystals he was able to produce diffraction patterns from X-rays and so to

establish the wave as the basic model for X-ray radiation. This cannot be regarded as a direct demonstration for it depends upon the acceptance of another model, the lattice model of crystals. Indeed in X-ray crystallography the situation is almost reversed for the wave-model of X-rays is now taken for granted and the structure of crystals investigated by means of the diffraction patterns they form.

2. The use of a model as a helpful device is illustrated particularly by those analogies that are specifically disclaimed to be representations of the ultimate structure of nature. For example in investigating drops use is constantly made of the idea of a balloon, as if the drop had a definite skin, and experiments are carried out to find just how far the parallel can be extended. Testing against the facts here is not aimed at showing whether or not there is a skin to a drop, but at determining how far the surface of a drop behaves as if it were a skin. This differs from (1) above in that the 'ballooness' of the drop is investigated, not through another model, but by direct investigation, while it differs from the 'candidate-for-reality' kind of model by the conscious reservation of the investigator that he is pursuing what is at best only an analogy.

III

The examples in Parts I and II of this chapter have shown, I hope, the way in which the facts lead us to generalizations and theories, and the way in which we investigate the appositeness and application of the descriptions and theories we already have by trying to test them by experiment. I should like to turn now to a brief discussion of what happens when all does not go well, when a descriptive generalization or more rarely a theory, is disconfirmed. This does not happen very frequently for as I have tried to show above generalizations and theories are built up in such a way that they always stand a good chance of being confirmed, a better chance of confirmation than disconfirmation.

A law of nature, such as Boyle's Law, is, our analysis has shown, the result of generalizing according to a standard procedure the descriptions of certain particular happenings. If the appropriate procedure of generalization is carefully and correctly applied then it is very unlikely that the generalization so formed will ever be shown to be *completely* wrong for just such a range of results as those upon which it was based. It is only when we begin to apply the law to cases further and further removed from those cases upon which it was based that we begin to run serious risk of a contrary case appearing and disconfirming the law. If a contrary case does occur there are three possibilities open to us.

1. The law can be upheld and the contrary case merely noted. This will be the appropriate action when there are various doubts in the investigator's mind about the control of extraneous factors in the experimental set up. To find an explanation for the contrary case in such circumstances we must investigate factors which are *not* included in the law to see if an unsuspected variation among them is not perhaps correlated with the contrary case of the law. Cases of this kind are usually fairly obvious. For example when a discorrelation between the tide and the position of the moon is found it is not usual to raise objections against the lunar theory of tidal action, but rather to look in the outside circumstances (configuration of the sea-coast) for some factor which will account for the anomaly.

2. If the disconfirming instances occur in conditions differing in some clearly definable way from the cases on which the descriptive generalization is originally based then there are two subalternatives.

(*a*) The generalization is upheld, but it is limited in its field of application and is always stated with an attached rider which sets out the limits within which it is a satisfactory description. For example when schoolboys learn Boyle's Law they learn that at constant temperature and for a given mass of gas the pressure of the gas is inversely proportional to its volume provided that the pressure is fairly low and the temperature fairly well removed from the critical point for the particular specimen on which experiments are to be carried out. The discovery of the limits of application of the laws of nature plays a great part in the development of a proper theory, for it is at the point of breakdown that the demand for an explanation becames most urgent. The fact that gases do not obey Boyle's Law at high pressures is a stimulus to further investigation for the explanation that the close proximity of the molecules of the gas one to another brings forces of mutual attraction into play practically demands that these forces should be investigated.

(*b*) An alternative that is frequently adopted in physics is the expedient of the postulation of an ideal substance. The law that is obtained in very favourable conditions is retained but is said to apply not to real substances but to an ideal substance which lacks those properties that · force us to apply limitations to the applications of the law. In a way the laws that describe the behaviour of an ideal substance serve also to define it, as that substance which obeys these laws. This expedient has been adopted to deal with the aberrations in the gas laws under extreme conditions, the physicist, especially the teacher of physics, distinguishing between the relatively simple ideal gas laws and the simplified substance whose behaviour they describe on the one hand, and real gases and the modified laws which describe their behaviour on the other. The whole subject of mechanics is like this, taking as its subject matter ideal bodies which lack certain of the awkward properties that make the laws of motion of real bodies so exceedingly complex, and so difficult to apply with any confidence.

3. When it is found, as in (2), that there are limits to the application of a generalization, that in different conditions from those in which the generalization was originally formulated there are disconfirming instances, attempts are usually made to formulate a new generalization which will include all the results obtained under all the sets of conditions that have been investigated. But the generalization so formed will be of such a nature that in the original conditions the original law will be derivable from the new, more general law. This can best be illustrated with a functional generalization. To continue the example of the gas laws: the general gas law in favourable conditions of pressure and temperature is

$$PV = RT$$

where P is the gas pressure, V the volume, T the absolute temperature and R a constant, the universal gas constant. This form of the law ceases to be effective for low temperatures and high pressures. An alternative that provides a much closer description at extremes of temperature and pressure is Van de Waal's equation:

$$\left(P + \frac{a}{V^2}\right)(V - b) = RT$$

the term b being interpreted as the total 'volume' of the molecules of the specimen and the term a/V^2 as the increase in pressure due to attraction between the molecules themselves and between the molecules and the walls of the vessel in which the gas is contained. Now in the conditions under which the original form of this law above was formulated the constants a and b are small in relation to the total volume occupied by the gas. It follows then that:

(i) $\dfrac{a}{V^2}$ tends to zero

(ii) $(V - b)$ tends to V.

Thus under the required conditions Van de Waal's equation leads back to the ideal form

$$PV = RT$$

This discussion of disconfirmation of descriptive generalizations might be summed up by contrasting the effect of a disconfirming instance in logic and in science. In logic a single contrary instance is sufficient to falsify the generalization from which it was derived, but in science the appearance of a contrary instance is not a signal for a rejection of the generalization but rather for more intensive investigations into the conditions under which the generalization holds. Scientists are the misers of logic who throw nothing away but rather try to find how to fit the restricted knowledge of their predecessors into the wider knowledge they themselves possess.

When a prediction made by means of a theory fails, there are still more alternatives to simple rejection than there are for descriptive generalizations. The failure of a theory is in the first instance the failure of a simple generalization, for it is only by the use of derived generalizations that a theory finds application to the world and is put in risk of disconfirmation. If either of the alternatives (1) or (2) above is adopted to deal with an aberrant case of a derived generalization then the theory in which the appropriate generalization finds a place is not rejected, but there now exists a stimulus towards its deployment or development. We may be able to pick upon the factors which are relevant but not taken account of in the law under investigation as in (1). On the other hand we may be able to develop our model in such a way that without any major change the aberrant cases are accounted for. The third possibility above is in fact what from the point of view of theory-construction I called the deployment of a theory, for the additional terms are added in accordance with the extended character of the model. It was originally a model built up of weightless, perfectly elastic point-masses and it is deployed by the addition of size and cohesiveness to the original properties of the elementary particles. My point is that as far as theory is concerned Van de Waal's equation is *consistent with* the general gas law.

But theories are discarded from time to time. There are two distinct kinds of case, corresponding to the differences between theories constructed by the use of models based upon analogies and theories constructed by the use of formal mathematical models. With the former kind serious failure to account for certain phenomena produces a casting around after new models, models which will give us a picture of the workings of reality in which the new phenomena can be included. But failure to account for new phenomena or the making of predictions which are falsified by experiment are not the only stimuli which produce the search for new models. There may be what might be called direct dissatisfaction with a model. This may be either because in the way described above a model may turn out not to be in fact the fine structure of the world; or because in the process of development the model has become of such great complexity that it is felt that something simpler must be invented and substituted for it.

There are then four major sources of dissatisfaction with theories based upon analogies and models.

1. Failure to account for all the phenomena which it might be expected to account for. The discovery of interference phenomena in optics destroyed the corpuscular theory of light at least at an elementary level, for the corpuscular model was simply incapable in the hands of the physicists of the eighteenth century of providing any sort of explanation of interference. The model was dropped and the wave-model substituted with the result that the new phenomena *as well as* those accounted for in terms of the old model could be expressed in the new theory.

2. A failure in power of prediction with an otherwise satisfactory model is, more often than not, counted as a minor failure and stimulates, not so much attempts to develop a new theory altogether but to bolster up the old by developing or deploying the model upon which it was based. When Bohr found that Rutherford's atomic model, as it stood, did not serve for the

correct prediction of the character of the spectrum of hydrogen, he did not discard it completely, after all Rutherford had produced very good reasons for adopting it, but he developed it by adding the subsidiary hypothesis that electrons were confined to certain definite orbits, so that in this new developed form the model was capable of being used to make correct predictions.

3. The example I used above to illustrate direct investigation of a model is also a good example of this method of disconfirmation for it is generally agreed that Fage and Townend succeeded in showing that there was no stagnant sub-layer at a phase-boundary. The result of this, together with certain other considerations, summed up by one of the investigators as 'commonsense grounds', is the rejection of the stagnant sub-layer as a model. A new model is then adopted such as that proposed by Professor Danckwerts[3], in which it is postulated that the surface of a liquid is not stagnant but that it is continuously being replaced with fresh liquid. A complete alternative theory of absorption has been developed on this basis. It is interesting to observe that there are no apparent differences in the predicative power of the two theories based upon the alternative models as has been demonstrated by Professor Garner[4] for a wide range of conditions. We must look to the direct observation of what the models postulate as the mechanism of absorption for reasons for the supercession of one by the other.

4. The use of simplicity as a criterion I shall discuss in the chapter which follows. The introduction of this criterion is based upon the need to take account of the complicating effect upon a theory of the further deployments and developments that are needed to account for the wider range of facts that continual experimenting provides. It marks a boundary between what might be called external criteria, which have been discussed in this chapter, and internal, non-factual criteria by which we make more detailed assessments of the satisfactoriness of theories.

Those theories which are developed in a purely formal way are rarities in science, but are nevertheless often very fundamental. The builders of the most comprehensive theories of the character of matter, unable to find satisfactory analogies to build upon, have taken the only step open to them and essayed the construction of formal systems which through the rarefied kind of analogy that exists between an algebraic function and the history of a system can serve as the basis for theory-construction. It has been pointed out by many logicians (including K. R. Popper[5] and R. B. Braithwaite[6]) that the failure of a single predictive instance in a tightly connected formalized theory has repercussions throughout the theory. This point has sometimes been put in the form of the epigram that in every experiment a whole science is put to the test.

We have seen that there is reason to believe that this is too extreme an expression of the effect of a disconfirming instance on a theory which is based upon a pictorial or mechanical analogy even when the theory is formalized. Nevertheless while patching-up is always an alternative to outright rejections there have been, even in theories that depend much upon models, experiments on whose satisfactory outcome the acceptability of a whole theory can be said to rest. A good example of this kind of experiment is that which was used by Cavendish to check the hypothesis that attraction and repulsion in electrostatics were expressible in an inverse square law. If we assume the general background of electrostatic theory it is possible to deduce that if the electric field strength is expressible as an inverse square law then there will be no resultant field within a charged sphere. The experimental testing of this prediction is extremely simple to arrange, requiring a charged sphere with a spherical conductor inside it, connected through a current-sensitive device, say a galvanometer, to earth. The experiment is judged successful in the absence of any reaction by the galvanometer to the charging of the outside sphere. It is not too much to say that in this experiment both the particular law and the general theory of electrostatics, particularly that part of the general theory that relates to electrostatic induction, are put to the test at once. The null result can be predicted from the inverse square law only within the general framework of electrostatic theory, while without the theory no such result could be predicted from the inverse square law alone.

Schematically this kind of testing can be represented as follows. If A, B, and C are the laws required to make a prediction a, then if a turns out to be correct we have the following possibilities.

A	assumed	confirmed	assumed
B	assumed	assumed	confirmed
C	confirmed	assumed	assumed

If a turns out to be incorrect then any one, or any pair, or all three may in fact have to be rejected, but from the single incorrect prediction it is not possible to say which. In terms of the example, though Cavendish thought of the experiment as confirming the inverse square law we could with equal justice look upon it as confirming the theory of electrostatic induction under the hypothesis of the correctness of the inverse square law.

However such laws as A, B and C may form some kind of hierarchy, say that A and B are premises from which C is deduced. Then failure of C as a rule of prediction *must* lead to the ramification of suspicion throughout the system, and in our simple example to doubts about A and B, for if they had been true so should C have been. As I suggested above, in fully formalized and strictly deductively related theories patching-up is fairly difficult for no part of the theory is wholly independent of the rest. It should not be supposed though, as some writers have tended to, that each time an advance is made it proceeds by a clean sweep[7]. In fact, when a serious difficulty, arising either from factual disconfirmation or through some internal failure, besets a formalized theory then it is necessary for the theoretician to begin again, but not from scratch, for there were good reasons for adopting the deeper hypotheses of the old theory. He looks first for modifications which ramifying through the whole deductively related structure will lead to predictions that are in accordance with the facts. I want to say that 'beginning again' is a relative matter, for it would be wrong to suppose that new beginnings are entirely independent of and unconnected with the old[8]. The search for an adequate atomic theory which has been proceeding under the pressure of experimental discoveries for something like fifty years, has involved, not the radical rejection of the unsuccessful and the building of the wholly novel, but the gradual modification of the old. What must be emphasized is that in a strictly formal theory modification of any basic postulates modifies every part of the theory.

REFERENCES

1. A. Fage and H. C. H. Townend, 'An examination of turbulent flow with an ultramicroscope', *Proc. Roy. Soc.*, A, **135** (1932).
2. T. S. Hanratty, 'Turbulent Exchange of Mass and Momentum with a Boundary', *Amer. Inst. Chem. Eng. J.*, **2**, p. 319 (1956).
3. P. V. Danckwerts, 'Significance of Liquid-film coefficients in Gas Absorption', *Ind. Eng. Chem.*, **43**, pp. 1460–6.
4. F. H. Garner, S. R. M. Ellis and D. C. Freshwater, 'The Comparison of

Vapour-Liquid contacting apparatus', *Trans. Inst. Chem. Eng.*, **35**, p. 61.
5. K. R. Popper, *The Logic of Scientific Discovery*, § 18.
6. R. B. Braithwaite, *Scientific Explanation*, Chaps. 2, 3.
7. T. S. Kuhn. *The Structure of Scientific Revolutions*.
8. Most of the points concerning the testing of theories made here have been restated in somewhat bizarre form by I. Lakatos, in his 'The Methodology of Scientific Research Programmes', *Philosophical Papers*, Cambridge University Press, Cambridge, Vol. I, 1978.

7 Non-factual Criteria

Cases of conflict, between alternative theories and even between alternative forms of description, often occur in science. For example in discussing the disconfirmation of theories in the last chapter I mentioned the current dispute between alternative theories of boundary absorption between which, at the moment, there are no ascertainable differences of fact, except the evidence of the direct investigations of Fage and Townend. As far as their power of predicting absorption rates is concerned either theory gives the correct results. How do we choose between them? Danckwerts in summing up his reasons for rejecting the stagnant film hypothesis says, very noncommittally, that commonsense grounds counsel its rejection, for it is not a reasonable model. It does not postulate the sort of thing we could reasonably be expected to accept as an underlying mechanism of transfer. This sort of judgement is extremely important from a practical point of view but must be passsed over by the logician for there is no question of giving a general account of 'what one could reasonably be expected to accept', for this is compounded of one's experience, one's previous knowledge, the state of development of the field in which one is working, and other imponderables of the kind. But there are other criteria, which may indeed form part of what Danckwerts means by commonsense grounds but which are recognized independently in giving judgement upon the satisfactoriness of a science and for which some kind of general treatment is possible. In discussing the same controversy as Danckwerts, Professor Garner advances the suggestion that the stagnant film hypothesis should be retained on the grounds that it leads to a simpler theory. There are many criteria of this kind, simplicity, economy, elegance and others and I shall attempt some explication of these non-factual criteria in this chapter.

I

It would be helpful to begin a discussion of simplicity by considering the basis upon which judgements of the simplicity of physical systems are made. Broadly speaking there are two different but related senses of 'simple', when it is used as a predicate of things.

1. We may, when we call something simple, mean that it is made up of very few parts., We use *economical*[1] as a synonym in this sense. An economic structure will be one in which the fewest possible parts have been used.

2. However, calling something simple may draw attention not so much to the paucity of elements in its construction but to the way it is put together, to its structure. There does not seem to be a synonym for 'simple' in this sense in common use, and I shall distinguish this kind of judgement as a judgement of structural simplicity.

In both these senses 'simple' is used in making judgements of descriptions and explanations. But in these matters a third sense occurs which often cuts across the other two, namely that when we call something, say an explanation, simple we mean no more than that it is familiar or even easily grasped. It is easily grasped just as something that is 'really' simple is easily grasped. In this sense 'simple' is often used by scientists to the annoyance and bafflement of laymen. There is no sense then in trying to give general criteria for making judgements of simplicity. Apart from the fact that 'simple' may be being employed in any of the above three senses, any criterion will depend very much upon the kind of description or explanation being judged. The significance of a simplicity judgement, and a choice of the simpler of two alternatives in general, can still be appreciated from a particular case. For clarity I shall choose as examples the more formal kinds of descriptions and explanations.

The operation of simplicity criteria in choosing a satisfactory description can be seen most easily in the case of descriptions in the form of functional generalizations. Three sets of criteria operate in selecting a simple functional generalization as our favoured description.

I. *The criterion of integral numbers.* Suppose that one were expressing the results of a set of experiments, say on Ohm's Law, and found that a perfect fit with the actual numerical values of observables derived from the experiments could be got with the expression

$$V^{1\cdot01} = IR$$

According to the criterion of integral numbers we express this result as

$$V = IR$$

choosing an integral index for V rather than an exact fit with the results of the set of experiments.

II. *The criterion of the lowest power.* In the example discussed in detail in the previous chapter, (p. 115), the facts showed a considerable experimental variation but it was possible to build a simple generalization. The principle behind the process was that if the numerical variations are small then high powers should be discarded from the function that is supposed to represent the results, and we should aim for representation by a function of the lowest possible algebraic power. In choosing a linear function rather than one to the power of eight an exact fit with experimental results was given up as it is in (I) above.

III. *The criterion of the reduction of observables.* Description must be economical, that is the least number of variables must be used in the function in which the description is expressed consistent with representation of the facts within the bounds of extreme experimental variation. This is a reflection in the formalism of the pressure towards a reduction of observables discussed in Chapter 3. Concrete expressions of the use of this criterion are the very general theories in physics, the cosmologies, in which, for example, attempts are made to find a single field expression from which the characteristics of the electric, magnetic and gravitational fields can be derived.

The ideally simple functional generalization then is characterised by having integers and preferably small integers for its numerical coefficients and indices, by being of a simple algebraic form, that is of the lowest power that will still give a reasonably close approximation to the results of experiments, and by being expressed with the fewest possible variables. The procedure by which in the example in Chapter 6

$$\frac{aV^8}{I^8} - \frac{bV^7}{I^7} \quad \ldots h = 0 \tag{1}$$

is converted into

$$V = IR \qquad (2)$$

satisfies at least the first two of these criteria. By eschewing any attempt at the representation with additional variables of the small variations from the exact match of calculation with experiment the conversion at the same time satisfies the third.

What is to be gained by adhering to these criteria? There are advantages to be derived both from the side of the eventual confrontation of predictions from the generalization with the facts and from the side of the rest of the science into which the generalization must finally be made to fit. It is often said by scientists that of any pair of alternative forms, other things being equal, the simpler has a higher prior probability, that is stands a better chance of being confirmed by successfully serving as a prediction rule. This is relatively easy to demonstrate, and is indeed obvious with our example of a derivation of Ohm's Law above. The expression (1) above is derived as an exact expression of the results of eight experiments with voltmeter and ammeter. It takes account of both positive and negative variations even if they are random variations. When we predict a new result by substituting new values for all but one of the variables in this equation we are without any guarantee in the method of formation of the generalization itself that by making such a substitution we will hit on exactly the value of the result of the appropriate experiment. And yet this form of generalization was built upon the principle of the *exact* representation of the results of experiment. On the other hand the form of expression (2) which in its formation averages out any randomness in the results does not pretend to give a result more accurately than within the range of the variation which it averages. Since there *are* random variations it follows that expression (2) has a much higher prior probability of being confirmed than has expression (1). Putting this in another way we can imagine expression (2), $V = IR$, represented graphically. Both the function and the corresponding graph will represent an averaging out of the variations which the additional terms in expression (1) were intended to represent. Since this is the case we may not expect another experiment to give a result which when plotted on the graph would lie exactly on the curve representing expression (2), any more than we would expect it to lie on the curve representing expression (1). However, the point of the latter is that it includes exactly the results already obtained, but of the former that it does not but rather represents their general trend. Since in any further experiment we can only expect a result in keeping with the general trend of the results so far obtained expression (2) will prove to be a more satisfactory representation of the concomitant variations that we are recording. A similar argument could be developed to show the advantage of choosing integral coefficients and indices.

Fitting a generalization into the body of a science can be broken down into two distinct kinds of assimilation. (i) It can be assimilated because it is found to fit into a deductively related system of similar generalizations; or (ii) it can be assimilated because it is found to bear some sort of analogy either formally with other laws of nature, or informally with some model. If generalizations are always constructed and selected with the idea of simplicity in mind then there will be more likelihood of finding for any given simple form other generalizations of the same form, and if economy has also been a guiding principle, using the same set of variables. Generalizations having these characteristics are clearly easier to fit into deductively related systems than general descriptive expressions of the type of (1) above. According to the same principle the simpler the generalization the more likely it is to have analogues both in the form of other established generalizations and of simple models. These criteria are never precisely formulated, nor should they be, for their vagueness is a reflection of the fact that rather than there being hard and fast rules for doing science there are general trends in scientists' methods and modes of thought that assist rather than determine the processes of generalization and theory building. The final arbitration must always be in terms of the facts.

In order to see where desirable simplicity characteristics are to be looked for in explanations the basic conditions for a minimally successful explanation must be considered. These, as was shown in detail in Chapter 1 and 4, are:

(i) All particular happenings in the field which the explanation covers must be able to be shown as 'consequences' of the theory, in the sense that given some starting point among particulars there must be a general statement or statements in the theory which will serve for the derivation of *any* new factual statements in the field from those which are given as premises.

(ii) The explanation must make us understand the phenomena

for which it is required. This requirement is satisfied by the
linking of a model, more familiar in character than the facts, to
them. When this is not possible, then we make do with a
formalism which connects up all the generalizations we have
made in the field under consideration into a deductively linked
system.

It is an aim of scientists to satisfy these requirements as simply
as possible.

(i) This aim is achieved for the derivation of new factual
statements by building up the simplest possible descriptive
generalizations. It is then ensured that the passage from premises
to conclusion will be as direct as possible since it will proceed
according to the simplest possible rule of prediction. There is then
no special requirement here not satisfied by our stating our
generalizations in the most desirable form. For example the
prediction of the current in a conductor at a certain voltage is a
good deal simpler process with

$$V = IR$$

than with

$$\frac{aV^8}{I^8} - \frac{bV^7}{I^7} \ldots - h = 0$$

If we want to go into detail, fewer operations are required, and
each operation, not requiring the raising to the eighth or seventh
power, etc. as in the second form, is itself simpler than the
operations which require these moves. If when we demand an
explanation we mean something like 'Why is the current 4 amps
when the voltage is 8 volts?', the answer, trivial in this case, is that
the resistance is 2 ohms and the relevant law is $V = IR$, then the
simpler the form of the law the simpler the explanation.

(ii) In assessing the simplicity of a model all three major criteria
of simplicity are needed, for while the ordinary needs of human
understanding require that anything, model or whatever it is, be
economical in concepts and simple in structure, the advance of
science through the deployment and development of the chosen
model will depend upon our familiarity with it. Both deployment
and development, the extension of the model without the addition
of new concepts and the grafting on to it of subsidiary models,
demand a complete familiarity with it. It would, for example, be
hopeless to pick upon balls in a box as a model for the behaviour of

gases if we were not thoroughly familiar with the behaviour of balls in a box. In many cases when a theory is judged to be simple attention is not being drawn to the paucity of concepts employed in its construction or to the simplicity of its structure but to the fact that the model which it is based upon is one which either the author of the theory or preferably everyone, is quite familiar.

In the discussion of induction in Chapter 5 I suggested that useful questions about induction resolved themselves into two independent sub-questions. These were (i) 'How is a hypothesis (be it generalization or theory) first discovered?' and (ii) 'How is it tested?' It was argued in that chapter that no general answer can be given to these questions but that there are as many cases as there are kinds of generalizations and kinds of theories in the sciences; and for each there is a great variety of procedures of both confirmation and discovery. Among these is the use of simplicity as the basis of methods both for the making of discoveries and for the assessment of hypothesis, particularly when the testing of a hypothesis involves the selection of one of a pair of alternatives.

A famous example of the use of simplicity as a criterion for the assessment of an alternative hypothesis was the compromise which terminated the disputes about the acceptability of a heliocentric hypothesis as an explanation of the movement of the planets. At the time when the disputes occurred there was little factual evidence upon which to base a decision as to the objective truth of either the geocentric or the heliocentric theory. Both were, in a very real sense, hypotheses. In order to avoid conflict with the Scripture some sort of compromise had to be reached, so that the heliocentric theory could at least be recognized officially as a hypothesis. The grounds that were finally found acceptable to both sides were that since there was no factual evidence upon which to base a decision between them the superior simplicity of the heliocentric theory* made it acceptable, but (and here the official arbiters of science were careful to stand their ground) as a method of calculation only. This was a very striking example, but there are many ordinary cases too. Decisions are being made every day in laboratories on the grounds of simplicity. Indeed it is a routine part of a scientist's daily task to reject everything but the simplest expression of his results that he can find. These rejections of course need not be formulated as conscious designs but they

---

* Not, be it noted, in the hands of Copernicus whose system was in the end structurally more complex than Ptolemy's.

determine the direction in which for example improvements in *ad hoc* hypotheses are sought.

It was argued on several occasions by Einstein that simplicity judgements were the basis of his method of discovery. The method he advocated has been summed up by one of his pupils as follows: 'Einstein's achievements seem to have been attained by means of exactly those methods which he described as the appropriate methods of theoretical physics: the physical world is represented as a four dimensional continuum, a Riemannian metric is adopted, and in looking for the "simplest" laws which such a metric can satisfy, he arrives at the relativistic theory of gravitation of empty space. Adopting in this space a vector field, or the anti-symmetric tensor field derived from it, and looking again for the "simplest" laws which such a field can satisfy he arrives at the Maxwell equations for free space. He also states that in the paucity of the mathematically existent field types, and of the relations between them, lies the hope of comprehending reality in all its depth'[2]. Putting this untechnically we are being recommended to adopt a method of discovery that is to be represented as follows: we adopt first some very general principle of explanation, in this case the notion of a field, and then search for the simplest possible detailed principles within this general picture. Einstein himself says[3] 'these fundamental concepts and postulates which cannot be further reduced logically form the essential part of a theory which reason cannot touch. It is the grand object of all theory to make these irreducible elements as simple and as few as possible, without having to renounce the adequate representation of any empirical content whatever'.

The claim that with this method we can 'comprehend reality in all its depth' should not be taken too seriously, for it depends upon an unprovable metaphysical principle which Einstein firmly held, that the structure of the world always tends to be as simple as possible; a kind of structural analogue of the Law of Least Action. The holding of this principle is, I think, a case of the fallacy of confusing one's model for the understanding of the world with the world itself. This fallacy has been committed by some very eminent scientists, indeed it is one to which they are peculiarly liable. For example it was argued by Sir James Jeans[4] that since the only adequate model for the behaviour of fundamental particles was a mathematical one, the world must be mathematical in character, a view which he summed up in the epigram 'God must be a mathematician'. Einstein in advocating simplicity criteria is no doubt describing a very powerful method of gaining an understanding of the world, but it is not to the world but to his explanation that the criteria of simplicity apply.

Also within the category of discovery-procedures comes the use of simplicity criteria in settling what sort of corrections are to be made to a science or perhaps a single theory when some serious fault is found with it. When a theory breaks down we may choose to invent a hypothetical entity which at the time of its introduction we have no way of observing, rather than make large-scale alterations in the theory, and this choice is often determined by considerations of simplicity. It happens that from time to time an unusual experimental result is obtained within a certain physical scheme. It is a result that is sufficiently wide of the hypotheses that should cover it to seem to require some additions to the theory. It then becomes a matter of choice in many cases in what way we are to alter the theory. There are in general three courses open to us. We may say that the result is outside the theory, so that while we note it as being unusual we continue to work with the original theoretical structure. Such a decision might be justified on the grounds that there could be some undiscovered fault in the apparatus that caused a change in a condition normally maintained constant. Alternatively we may decide to alter the hypothesis under which we include the experiment that has produced this aberrant result. Unfortunately in a sufficiently highly developed science we cannot do this and leave the other hypotheses of the science untouched, so that this course will be undertaken with the greatest reluctance. The third possibility is to take the simplest course and to save the theoretical structure by introducing a hypothetical entity. It will be, initially at any rate, an unobservable that is postulated to behave in such a way that the original hypothesis which it was introduced to save, covers its behaviour, and in so doing accounts for the anomalous result that led to these changes being made.

Logically speaking there are two different cases here, which I shall mark by distinguishing two different kinds of postulates: (i) those introducing a hypothetical but particular *object* of a kind with which we are familiar, (ii) those introducing a *class* of entities of an altogether new kind. As an example of the former; it was found that the orbits of the planets did not have exactly the shape required by theory. The alternatives were either to alter the planetary laws or to postulate a hypothetical object, an unobserved planet. The latter alternative required no changes in the basic scheme of the laws for by hypothesis the new planet's behaviour is predictable from them equally with those which had at the time been observed. More powerful telescopes allowed the hypothetical planet actually to be observed. Clearly from the point of view of theory it was simpler to introduce the new object than to change the laws. As an example of the introduction of a hypothetical kind of object we can take the introduction of the neutrino. In certain atomic disintegrations the masses and velocities of the particles experimentally observed was insufficient to account for the momentum relations demanded by theory. The question then arose as to whether it was simpler to alter the law of conservation of momentum for sub-atomic particles and so develop yet another mechanics or to postulate a class of theoretical entities to account for the discrepancy. These would be unobserved and would be supposed to be projected in the disintegration with just the right amount of momentum and in just the right direction to account for the discrepancy among the observable particles. It was finally decided to invent a particle with these properties, the neutrino[5].

II

Another important class of theory-appraisals which do not depend upon factual criteria are those made with expressions from the vocabulary we normally use for expressing aesthetic judgements. Since the full range of this vocabulary is not used by scientists I shall call such judgements the expression of *quasi-aesthetic appraisals*. For example though we use 'elegant', 'pretty', 'neat' and sometimes 'beautiful' we do not use 'charming', 'delightful', 'lovely' or 'handsome'. Quasi-aesthetic disapproval is marked by the use of expressions like 'clumsy, 'ugly', 'dull', etc. The commonest expression used in this way is 'elegant' and I shall try to give an account of its use.

As with most of the judgements we make of theories, the point of calling some construction elegant can most easily be seen in cases where the construction is formalized. Mathematicians, for example, commonly distinguish proofs according to their elegance. If we want to see why a fairly specific term like 'elegant' is chosen from the normal asethetic-judgement vocabulary to express mathematicians' quasi-aesthetic appraisals we must try to see what features of difference amongst proofs are picked out by the use of this word. One proof differs from another in the same field both in length (number of steps) and in form (complexity of steps), and in whether or not it makes use of some ingenious device. A proof may be called elegant because

(i) It is short, that is has few steps.
(ii) The steps themselves are uncomplicated.
(iii) It makes use of an ingenious device.

Now the point of using an ingenious device is that it should lead to either or both of conditions (i) and (ii) being fulfilled, so that we might be tempted to assimilate (iii) to (i) and (ii). It might seem then that a judgement of elegance is just an obscure way of commending the simplicity of a proof. But the usual commendatory judgement of a proof is expressed by using both 'simple' and 'elegant'. It would seem then that something more is being said by the use of 'elegant' than can be said by the use of 'simple' alone.

To understand the point of using 'elegant' we must turn to its root-meaning. It is derived from the Latin *elegans*, meaning 'chosen skilfully or carefully'. We can call a proof elegant as well as simple *when skill has been shown in choosing a simple proof*. It is the exercise of this skill and not just the simplicity of construction that is the occasion for quasi-aesthetic satisfaction. That is to say the proof to be judged elegant must achieve its effect, the proof of a conclusion, (something satisfying in itself) in a specially satisfactory way. Quasi-aesthetic satisfaction is an additional satisfaction, and we can only introduce mention of this additional satisfaction in the case where a less satisfactory procedure is known or suspected by us to exist. The simplicity of a proof is judged by contrasting it with all other proofs, while elegance can only be judged by contrasting a proof with another which does exactly the same primary job. Quasi-aesthetic satisfaction is the additional satisfaction that comes from knowing that there are less satisfactory ways of proving something than that which we have just produced. A common form of expression for a quasi-aesthetic judgement is 'in this proof certain inelegancies have been avoided', a form of expression which suggests that elegance is judged against a background of less satisfactory methods.

Other quasi-aesthetic terms that are commonly used are

'pretty' and 'beautiful'. I shall not discuss these in detail. 'Pretty' is used (with 'neat') to mark the effectiveness of the kind of device whose introduction into a proof would lead to those economies that call for a judgement of elegance. There is a similar use of 'pretty' derived from fencing, in the judgement 'a pretty stroke'. 'Beautiful' marks the unspecific commendatory appraisal, so that like 'good' further details and a setting for its use must always be described. I would like to say that to call a theory or a proof beautiful is to say that it is particularly satisfying but it is to leave open, in a way that the use of 'elegant' does not leave open, the question of what it is that is satisfying in the proof.

III

Logicians have often used the words 'fruitful' and 'fruitfulness' to mark differences in satisfactoriness not only among individual laws and theories, but among the concepts in which these laws and theories are expressed. An examination of the working journals of scientists suggests that these two words are seldom if ever used in practice but nevertheless one must grant, I think, that something that is perhaps best expressed by these words is included in many judgements of satisfactoriness. I believe that it is this notion of fruitfulness that scientists are aiming at expressing when they speak of a generalization, a theory or a concept as being useful, or as being suggestive. Fruitful theories (employing fruitful concepts) are those which are effective in leading both to new facts through successful predictions and to new generalizations through the extension of the theory in one of the many standard ways. In fact a fruitful generalization or theory is one to which one of the methods of development which have been described in earlier chapters can be applied and which then yields hypothetical facts, laws, and theories, which can be tested, judged and approved by the criteria which I have in these last three chapters been describing as the basis of confirmation-procedures.

Fruitfulness, then can be looked upon as a criterion for making selections among the more or less fruitful, but this is not so simple a process as it seems for fruitfulness is not a property that can be ascribed to a generalization by just looking at it. It can only be ascribed to a generalization when it has been put to work say as a prediction rule. When the new facts that have been predicted by means of it have been tested and the predictions found correct then we can call the generalization fruitful. The other non-factual criteria that I have described have been applicable to the sciences as static systems, but this criterion can only be applied to a science in action. We might say that the fruitfulness of the various elements in a science is a measure of how far, and how successfully, they contribute to the realization of the general aims of the science.

The most important question then in understanding how selections can be made on the basis of fruitfulness is the general one of the aims of science. From an anthropological point of view these aims are those described by Francis Bacon, Power and the Understanding that augments that Power, but this is not an answer to the logical problem. Translating these demands into logical features of the sciences has been in part the aim of this book. It seems that what we need to fulfil these aims are:

(i) A complete system of rules for making predictions. That these should be fully satisfactory for their purpose can be ensured only by basing them upon a complete system of well-grounded generalizations in which the world is accurately described.

(ii) A fully coherent system of laws. The generalization-prediction rules which are the laws of nature should be expressed in a system where all are logical consequences of a small number of principles.

(iii) Finally, it is desirable to have, though we can at a pinch do without, an all-embracing picture of the workings of the world. This will consist in a description, whose component statements parallel the statements in the system of laws, of a model or connected set of models which match the world, and whose imaginary workings issue in the same consequences as the unknowable mechanism of Nature.

A fruitful concept is one which when used in the expression of facts or laws leads towards the fulfilment of these aims, a fruitful generalization is one which either by virtue of ingenious form or of original content subserves these purposes, and a fruitful theory is one whose construction and character is such that it promotes these aims.

REFERENCES

1. E. Mach, *Scientific Lectures*, p. 190 ff.
2. M. Rosenthal-Schneider, *Einstein; Philosopher-Scientist*, p. 137.
3. A. Einstein, *The World as I see it*, p. 134.
4. J. Jeans, *Physics and Philosophy*, p. 16.
5. For a readable survey of simplicity considerations see E. Sober, *Simplicity*, Clarendon Press, Oxford, 1975.

8 Postscript to Second Edition

Up to this point in the analysis of scientific activity I have been working within the system of concepts that most scientists subscribe to in their own reflections on their ways of thinking and working. It is now time to step outside this system since it does not have the objectivity and universality that is sometimes assumed by those who use it. In the discussion of theories and the relation of evidence to hypotheses I have presumed that the logical distinctions between facts and theories and between hypotheses and data are aspects of one deep lying division. On the one hand there are the speculative constructions of the scientific mind, laws and theories; while on the other are the indubitable facts which form the bedrock of scientific knowledge, and which we can properly call 'data', that which is given.

We have already noticed how the raw material of experiment and observation is tidied up for use as data. Now we need to look more closely into how that tidying up is achieved and what are the influences that bear upon it. The first stage of analysis will show the extent to which theory and fact interact with one another. We shall briefly examine the extrapolation of this interaction into the idea that each theory defines a distinctive world and a unique system of concepts cut off in some fundamental way from its successors and predecessors in the history of the scientific study of a field of interest. At this stage of analysis reference is confined to matters that are within the products of scientific activity itself, namely theories, concepts, laws and so on. A further stage of analysis of scientific assessment of theories and of the sources of the content of theories has been proposed. In this second stage the social relationships among scientists and of scientists to the larger social milieux of their times are offered as the explanatory basis for understanding why certain sorts of theories are invented and why this or that innovation is accepted or rejected. We will briefly examine the generalization of the idea that there are social influences on scientists to the exciting but extravagant doctrine that everything that scientists think and do is capable of sociological explanation without remainder.

In this chapter I shall be showing the limits of the mode of thinking I have described in the bulk of this text, which is the mode of thinking scientists widely believe themselves to use. I shall be trying to identify the mistakes that have led some commentators on scientific method to make the wilder generalizations of their limited but interesting insights. The main line of argument will be based on the extent to which what are taken to be the facts depend not on simple sensory experience, but on the theories, implicit and explicit, to which people subscribe. It is characteristic of philosophers to notice some hitherto overlooked feature of some mode of thinking, a feature which puts some sort of limit on the confidence with which that mode can be used to draw conclusions or to support a policy, and then to generalize their discovery so that they surprise us by drawing the conclusion that some commonplace activity is impossible. In the case in point it has been claimed that the scientists of one era are literally unable to understand the science of another era, since they do not believe the same theories as their predecessors.

The thought that theories and facts are not independent of one another goes back at least to the early nineteenth century writings of William Whewell[1]. As he puts it 'The Antithesis of Theory and Fact implies the fundamental Antithesis of Thoughts and Things; for a Theory . . . may be described as a Thought which is contemplated distinct from Things and are seen to agree with them; while a Fact is a combination of our Thoughts with Things in so complete agreement that we do not regard them as separate.' According to Whewell 'an activity of the mind, and an activity according to certain Ideas, is requisite in all our knowledge of external objects'. He goes on to emphasize that 'the Sciences are not figuratively, but really Interpretations of Nature'. These correlated antitheses are not fundamental. Behind them lies a kind of identity. 'The most recondite Theories when firmly established are Facts: the simplest Facts involve something of the nature of Theory.'

Some more recent writers have been concerned to emphasize the interrelation between theories and the meanings of descriptive vocabularies. This interaction has been dubbed the 'theory ladenness of descriptive terms'. When a physical scientist offers as a datum a certain reading of a meter, say an ammeter, his statement 'The ammeter reads 2.52 amps' cannot be fully understood in terms only of his physical sensations as he observes the location of the needle on the scale of the meter. The concept 'amp' is part of electrical theory, and is unintelligible without that theory. If he goes further and announces 'The current is 2.52 amps' the influence of electrical theory on his statement is still more obvious in that the term 'current' embodies a sketch or hint of an explanatory scheme in terms of which the behaviour of the whole set-up of which the meter is only a part, could be understood. It should also be noticed that the meaning of the term 'current' will change with the electrical theory which serves to give it meaning. If it is embedded in a theory of electrical fluids it can be read almost literally, but as part of an energy theory say, it has to be given a somewhat different reading. What then becomes of the 'fact' that the current is 2.52 amps? It will surely mean one thing in a fluid theory and something quite other in an energy theory. Is there then a common fact, one that is *the same* for both theories, and against which they might be tested by some kind of crucial experiment?

Well surely the behaviour of the meter is a common fact? The needle will reach a certain point on the scale regardless of the particular theory entertained by he who reads it, and so at least in principle, the possibility of finding such a 'common fact' that is in accordance with one theory and contradicts the predictions of another is not ruled out. This particular example has not been much discussed, but cases where even the idea of a common observational fact is hard to defend have been proposed. N. R. Hanson offered several such cases[2]. The most interesting and perhaps the most telling comes from the history of sub-atomic physics. Electrons (negatively charged particles) follow a characteristic track in magnetic fields. The question arose as to whether there were any particles of the same mass but positively charged. Anderson had found a track of a particle the length of which strongly suggested that it was of electronic dimensions. Was it a negative electron that picked up energy in the magnetic field (having a track of lesser curvature at the end of its flight) or was it a positive electron which lost energy and so curved more sharply under the influence of the field? Anderson chose the latter interpretation and so 'read' the track as that of a positively charged electron losing energy. If the track was made by an electron of negative charge the photograph represented one fact, if it had been made by a positron the photograph represented quite another. At the level of interpretation at which this matter entered into physics, say counted for or against Dirac's theoretical analysis, there were clearly *two* facts not one.

But one might object, surely the photograph is an object common to both theoretical frameworks, a common object which offers the possibility of different interpretations? But to accept that the photograph is indeed a *picture* of the track of an ionizing particle a very great deal of theory must be presupposed, and not only physical but chemical theory too. Only relative to a whole cluster of theories which everyone involved shared does the photograph exist as a collection of data at all. For someone completely outside the scientific community of the nineteen twenties and thirties the photograph has no object meaning at all. The idea then that there is one basic fact and that there are two theoretically influenced interpretations depends only on our not going deep enough into the degree to which the community capable of reading the photograph that way shared a common cluster of theories.

One might conclude from the considerations so far advanced that there was no objective basis for a common judgement of factuality at all. Yet that would be too extreme a conclusion. There are no basic facts, that is facts independent of any theory. But the argument admits of the relativisation of facts to more and more widely held theories, till we reach the kind of common sense theory which hardly seems like a theory at all. By that I mean the consensual beliefs we all seem to have that there is a material world changing through time, that there are other people, and we believe in the causal potency of certain things and events. A very general conceptual system, of which the above notions are the prime ingredients, seems to be the common possession of most of mankind, and to have stood the test of time, as the system of beliefs in terms of which commonplace practical activities necessary to the support of life have been conducted. So though one must admit that there are no common facts at the level of sophistication at which we recognize something as a photograph of the ionization trail of a sub-atomic particle, there are common facts at the level at which we recognize something as a persisting thing. The whole superstructure of beliefs may change yet a measure of agreement persists through it all. A final argument for the claim that the human perceptual apparatus does pick out some of the persisting 'furniture of the world' comes from the thought that the perceptual apparatus is the product of a long process of organic evolution, and so has been selected as at least practically competent to guide the actions of human beings and other animals in the world[3].

But what does this show about the basis of scientific knowledge? While the above considerations dispose of the extreme relativism espoused by some philosophers, they do not serve to justify any kind of foundationalism in the epistemology of science. This is because at the stage at which an observation or experimental result becomes a matter of interest to a scientist it has already been incorporated into a system of beliefs. Only as so incorporated could it stand in any kind of logical relation to a theory or hypothesis of which it might be thought to be a test. It is no use to science to be reminded that an ammeter is a thing. Its behaviour is relevant to science only in so far as it is interpreted as a device for measuring current. Darwin's observations of plants and animals are relevant to the theory of organic evolution only in so far as he came to see them, and to present them to his readers, within a framework of concepts that enabled him to express them as exemplifying family relations, that is relations of descent and common ancestry. But that is already to have transformed his observations by reference to a theory.

If the 'facts' are so intimately involved with the conceptual system and the corpus of beliefs that form the theoretical framework of a science, how can these facts serve to test theories? Tests cannot be simple applications of logical rules such as *modus tollens*, that is the rule that if p entails q and q is false, then p must be false. We need another notion. We might call it 'intransigence'. It may be that the 'matters' in the world, perceived in the first instance with the help of the common, universal but overly general conceptual system of commonplace action and belief, will not admit of reconstitution in the terms proposed by the theory. The elegant idea that one half of the cerebral cortex is concerned with linear linguistic performances, and the other with multidimensional more iconic forms of thought, was used to organize a great deal of information about human performances, in particular of people with damage to one or other hemisphere. But it has emerged that there are some features of brain function and of human performance, made into facts relative to a conceptual system more general than the right and left (bicameral) theory, which are recalcitrant, which do not admit of interpretation within that simple dichotomous scheme. To say that the bicameral theory has been falsified distorts the way it was tested. Rather it has been tried and found wanting, in certain matters. It will not do to conclude that because facticity is relative to theory that theories serve themselves by creating facts to suit themselves. This would be to miss two points: theories come in hierarchies of relative generality, and facticity may be sustained at the more general level for some matter which proved recalcitrant at another; a theory may simply fail to incorporate some matter. The world may be intransigent at that point. It just may not admit of being organized in accordance with the scheme.

Since theories and belief systems generally determine the meaning of even descriptive terms in the vocabularly scientists use to record their observations and experimental results, it has been suggested that a gulf of incomprehension must separate the believers in one conceptual system and its attendant theories from those of another. It has even been proposed that the scientists of on era literally cannot understand the writings of those of another, since the meanings of the latter depend on a theory which the former do not share[4]. Again we have a familiar pattern of philosophical reasoning. Facts are not wholly independent of theories, and meanings are determined, at least in part by the theories one holds, and not just by the things and events in the world to which one's terms refer. Therefore meanings are wholly internal to a system of thought. There are two reasons why the extreme conclusion must be rejected.

It is of course possible to learn the theories and to at least vicariously enter into the belief systems of another epoch. In this way one can come to grasp the meaning of their descriptive terms. One may never be sure that one has fully grasped the nuances of an alien system of thought but it can hardly be denied that one can get a foothold in it. Because a translation cannot be known to be perfect it does not follow that no translation is possible at all. But there is another and more fundamental argument. If we regard the meaning of a descriptive term to be determined by some set of rules of use, it seems reasonable to suppose that not all the rules of use for a term change at any time, through any one revision or replacement of theory, even though at the end of a longish time all will have changed. D. Papineau has shown how a kind of continuity can be established for mechanical concepts such as *vis viva* (an ancestor of both momentum and energy) through time by a careful historical study of the rules of use of a sequence of slowly changing concepts[5]. At no time is there a complete break with the rules governing the past uses of a term in the series, but every rule is modified in the end. We thus have the familiar phenomenon of continuity with partial identity, A somewhat different argument has been proposed by H. Putnam[6]. It depends not so much on a direct comparison of meanings but on the idea of conservation of reference. Though the gene concept has changed a good deal over the last eighty years we would be right to insist that the *same thing* was being referred to throughout its history; but we have had changing views as to what it is. Of course 'same thing' here is a colloquial way of talking about a type of thing. In a similar vein one could argue that the same thing was referred to by the term 'Terra Australis' as we now refer to with the term 'Australia', but our beliefs about the southern continent have been much modified.

Meanings and beliefs are the currency in which we discuss the content of theories. But there are also standards of reasoning, modes of theorizing, rules of evidence and the like which enter into saying what a science is. The system I have been describing in the chapters that have gone before is that in current use by scientists of more or less this generation. It is not hard to show that a very similar set of experimental and cognitive procedures have been in use for a very considerable time. Standards of accuracy and precision have been tightened up, but the general method of theorizing through an interplay between models and mathematics has been remarkably stable. Despite this apparent continuity and stability of method some philosophers have argued for a relativism of method as well as a relativism of the meaning and content of science. Toulmin, for example, has written of 'institutions of rationality' with the clear implication that like other institutions these have changed in historical times[7]. He distinguishes between the procedures of formal logic, universal tools to be used where appropriate, and some more general basis of rationality, in the light of which particular historically situated procedures that scientists have actually used, can be seen to have some virtue we might call 'rationality', as for instance the 'proper' things to do.

It is only the latter which is supposed to be historically relative and open. But this is, itself, a historical judgement. If it could be shown that some interconnected cluster of practical and cognitive techniques had time after time, subserved the aim of revealing more accurately what happened in nature and why it happened, that is by what productive processes, then the problem of rationality of science in this broader sense would be solved without the excesses of rationalism. Some who have argued along the same lines as Toulmin have followed the path of rejecting the simplistic methodological recommendations of Popper's early work in particular, based as they were on a generalization of the approved methods of formal reasoning, have then rushed off to deny that there are methods at all. (Cf. P. K. Feyerabend, reference 4, for an extravagant version of this argument.) It does not follow that because conjecture and refutation is not a capsule formula for scientific method, that another more sophisticated form of prescription cannot be sustained. I believe that evidence can be found to show that the scheme outlined in the earlier chapters of this work has been in general use by the scientific community at least since the time of Archimedes and Aristotle[8].

The gap of comprehension to which I have referred has been used by some philosophers as the basis of an argument to support a general scepticism about the possibility of defining any clear and universally applicable notion of scientific progress. Theories which draw on distinctive belief-systems and so serve to create descriptive terms whose meanings are as distinctive, are said to be 'incommensurable', literally they cannot be compared for any desirable epistemological properties such as greater accuracy or deeper insight into the processes of nature, properties whose comparison would ground judgements of scientific progress. A coherent cluster of beliefs and empirical methods appropriate to those beliefs was called 'a paradigm' by T. S. Kuhn[9]. Unfortunately his use of his own technical term was rather loose. As well as the use I have just described he also used it to refer to a specific piece of scientific work which served as an exemplar for research and which particularly clearly exemplified the cluster of beliefs and methods, the 'paradigm' in the first sense. The history of science was thought to be a sequence of closed episodes, as one paradigm succeeded another. Comparative judgements of theories and experiments and so on 'across' paradigms were thought to be impossible since a theory from one paradigm would be incommensurable with one from another, even if they seemed to be concerned with the same subject matter. Arguments against the idea that there are truly gaps in comprehensibility, particularly arguments which throw doubt on the idea that descriptive terms are wholly defined in terms of the theories to which they are related, also serve to undermine the general thesis of incommensurability[10].

I have now dealt, at least in broad outline, with the issues that arise from the realization that facts and the vocabulary in which we can describe them are intimately bound up with our systems of belief, some crystallized into well defined theories. But there is another way in which the objectivity of science has come under attack. The assault has come from the direction of the sociology of knowledge. It has been suggested that there are social forces at work in determining both the content and the assessment of theories even in the most highly refined sciences. Claims for the sociology of knowledge have ranged from the idea that there is a kind of general pressure on scientists to think in accordance with the main features of their social milieu to the extreme thesis that scientific activity and its products can be explained wholly in terms of the social factors involved in that production. I turn now to examine various suggestions about the social forces that influence the creation and recognition of knowledge.

II

In this section scientific work will be studied as an activity performed by a group of people who form an autonomous society—the community of scientists. The claim to see the scientific community as a society is based upon some obvious features of that community, the fact that it is clearly hierarchical in character, that it has a social organization, that it operates with an autonomous theory of valuation in which it determines the worth of the goods it produces, namely the products of scientific activity, so we can even say it has a kind of economy—all these things go to make us think of the scientific community as a society of some sort. Now this becomes of philosophical interest when the valuation of scientific products begins to be seen in sociological rather than epistemological terms, when notions like truth and knowledge become related to the social practices by which people make decisions whether to accept or reject theories, for instance. The early discussions of the social influences on the work of scientists were on the whole rather simplistic. T. S. Kuhn, for instance, noticed that the force of at least some of the criteria for the assessment of theories were best explained by the social position of those who promulgated them rather than by their intrinsic epistemological value. It was, for instance, pointed out that what counted as evidence was relative to the theories that were in vogue and the fact that a theory was in vogue was in part a matter of the social position of those who believed it. In recent years, the position has been stated much more firmly and much more generally. It has been argued that the concepts of scientific truth and scientific knowledge are not epistemological at all, but rather are capable of being explicated, at least in the way they are used in the scientific community, wholly in terms of social relations and social practices. If this argument were successful it would be devastating in its effect. If what counts as true and what counts as knowledge are social constructions of some sort, traditional philosophy of science is the pursuit of a mere illusion. How is such a dramatically different theory of science to be justified?

To examine this idea we must look quite closely at what sorts of analogue have been offered for scientific procedures by those who wish to see the communal facts of scientific life as dominant in what used to be thought to be the province of philosophy. As in any piece of scientific work, and the sociology of science is no exception, the heart of the scientific theory is the analogues which it brings to bear on the opaque and complex reality which it is attempting to understand. Economic models have almost wholly determined the way sociological analysis has developed. Scientific activity is looked upon as the manufacture of scientific artefacts, roughly speaking published papers[11]. It is the conditions under which they are produced and valued which provides the ground for the argument. Several different economic models have been proposed. The least plausible asks us to think of the production of scientific goods, namely published papers, as a kind of market economy. The valuation of the papers is not determined by their producers but by supply and demand. This idea seems to be inadequate not least because of the difficulty of identifying anything like a market for such an economy. There have been many studies of the readership of scientific papers, and the general conclusion seems to be that very few people read any given scientific paper. It has been claimed that, apart from the author and the referees, any scientific paper draws on average fewer than two readers. Perhaps the idea of comparing a scientific community to a guild economy would be more fruitful[12]. The medieval practice of organizing craftsmen into guilds, say the silversmiths or clothworkers, has some obvious similarities to the way many scientific communities are formed. The guild was sharply socially structured and the valuation of the items produced by guild members was determined by the guildmasters themselves. Not only did they set the quality, by and large they also set the price. So that it is social relations within the guild which are important determinants of what is accepted as 'good work' and the value put upon it. Ravetz has argued that the way scientific work is evaluated by journal referees, doctoral examiners science research councils and so on, has much in common with the system of valuation used by guilds. The scientific community selects its own senior members as the judges of quality. While this analogy throws some light on the actual process of evaluation it does not justify the strong claim of some sociologists of knowledge to replace epistemology. The leaders of the guild may have very good reasons for accepting or rejecting scientific work, reasons which could be given a traditional philosophical justification, for instance referees may have a practice of carefully scrutinizing the logical relations between cited evidence and allegedly well-supported hypotheses. Only those persons who subscribe to this practice may rise in the scientific social hierarchy to become referees and examiners. So, as a matter of fact, social prestige and power are associated with the practice of scrutinizing evidence. But clearly this relation is not constitutive of the scientific merit of closely checking the evidence.

A more subtle form of reduction of epistemology to sociology has recently appeared, associated with an anthropological stance in microsociology of science.

Recently a more strongly reductive theory of the relation of scientific practice to social influences has appeared. The mediating concept is that of 'interests'. What scientists do, as scientists, is supposed to be explained by reference to their interests, the implication being that these are social interests. The idea of trying to explain not only the choice of research topic but also the content of scientific theories and the favourable or unfavourable judgements made of them by reference only to social interests has been called the 'strong programme'. It has taken two forms. The 'Edinburgh' school refer to social interests defined relative to the structures of lay society, such as 'class interests', while the 'Paris' school refer to interests reflecting the inner workings of the scientific community itself, conceived as an autonomous social order.

The Edinburgh School

Characteristic of this school is D. Bloor's attempt to explain the development in the concept of number in the seventeenth century[13]. According to Bloor, Simon Stevin 'does not appear to have adopted the idea that "1" was a number because of the arguments he adduced . . . which were after-the-fact defences for a position which seemed quite evident'. The reason, according to Bloor, why it seemed natural that the 'unit' was also a number had to do with Stevin's location in the socio-economic order as an engineer. 'Those who opposed the new conceptions' constituted an obscurantist party, for whom number had a theological significance. At most Bloor establishes that some sort of loose correlation held between an interest in the 'new' mathematics and the needs of a new technology. But Bloor claims the needs of technology 'provide the most plausible *cause* of the change' (*op. cit.*, p. 105, my italics). However correlations only count as causes if some sketch of the mechanism or process by which their efficacy is mediated to produce the supposed effect. And we are offered no hint as to what that mechanism might be.

This example brings out the essential weakness in the Edinburgh school's version of the 'strong programme'. Their claim must imply, so far as I can see, that there is a community-wide self-deception in that the reasons scientists cite for accepting a theory, such as the assembly of compelling evidence, are superficial, and the real causes of their favourable or unfavourable assessments have to do with such matters as their class interest.

The philosophy of science implicit in this theory is some sort of conventionalism. The Edinburgh school choose as their philosophical foundation the Wittgensteinean idea of the language-game. They conceive of scientific practice as the setting up of different language games, sets of conventions for talking within a certain community and relative to a certain context.

These conventions are *somehow* determined by the interests of the speakers. The recent work of Barnes[14] involves a systematic exposition of a hermeneutical approach to analyzing scientific discourse. 'Conceptual fabrics', says Barnes, 'including those in the natural sciences, have the character of hermeneutical systems.' In particular the range of application of such systems involve 'successions of on the spot judgements' as to how to apply a concept. It seems to be implied that to understand the vocabulary of a science we would have to work our way back through the particular circumstances of each semantic decision. But even if this were so it would not rule out the possibility of a community-wide core of rules of use to which everyone adheres. Hermeneutics is invoked to try to defend the conclusion that the structure of the world is *wholly* man-made. Any less extreme conclusion would let in a measure of scientific realism, against which a philosophical justification rather than a sociological explanation of a scientific practice could be developed.

One further point of distinction concerns the relation of the views of the Edinburgh school to Marxism. It could be called a quasi-Marxist theory. The social facts which are supposed to cause the corpus of knowledge and the community's assessments of truth and falsity are essentially super-structural facts. For example they are matters of religious affiliation or rank and status in an educational system. If there are any base structure relations involved they are in general ignored by the Edinburgh school. It is noteworthy that Bloor relates the rise of the new mathematics to the *profession* of engineer and to the *belief* in natural theology and so on of those who opposed it. It is important to emphasize that the Edinburgh school uses the notions of truth and knowledge in the ordinary sense, which relative to the reductive force of the arguments of the Paris school, one might even call 'pre-critical'. Knowledge consists of certified true beliefs and truth and falsity mark the same sorts of appraisals as they have always been thought to. The Edinburgh school is concerned to provide a causal matrix to account for the genesis of belief systems.

According to Bloor (*op. cit.*, p. 66) the causally efficacious source of knowledge is 'social ideology'. 'The social ideologies are so pervasive that they are an obvious explanation of why our concepts have the structures they do.' But pervasiveness is no proof of causal potency. One could argue, *on the basis of the same evidence*, that it is because our concepts have the structure they do that 'social ideologies' reflecting them are so pervasive. A structuralist would argue that way. Further one can invoke Ockham's Razor against the causal usage of 'ideology'. Surely the idea of sexuality of plants, the concepts of crystalline structure applied to metals, the notion of endogamy etc. can be explained by reference to what can be observed about the way plants reproduce, about the distribution of high spots on a field ion micrograph of hot tungsten tips, and about the marriage practices of tribal moities. To invoke 'social ideologies' as well is to multiply entities without necessity. The Edinburgh school has not, to my mind, provided a convincing example of the social causation of knowledge, in the sense that only a social cause could be invoked. The difficulty of so doing is perhaps brought out in the attempt by Barnes to develop a method for proving that a particular interest engendered a particular piece of knowledge.

There simply is not, at the present time, any explicit, objective set of rules or procedures by which the influence of concealed interests upon thought and belief can be established, However, it remains possible in many instances to identify the operation of concealed interests by a subjective, experimental approach. Where an actor gives a legitimating account of his adhesion to a belief or set of beliefs we can test that account in the laboratory of our own consciousness. Adopting the cultural orientation of the actor, programming ourselves with his programmes, we can assess what plausibility the beliefs possess for us. In so far as our cognitive proclivities can be taken as the same as those of the actor, our assessment is evidence of the authenticity of his account[15].

Even if Barnes were able successfully to simulate the covert psychological process by which his scientist was the mere creature of his interests, would this prove anything of interest to philosophy? We must now distinguish between the project of reducing theory of science to sociology and the rather less interesting project of giving an account of the particular origins of a theory as a thought in somebody's head. To fail to make this distinction is to commit the 'genetic fallacy'. In this case this would mean that one presumed that because one had a psychological or sociological explanation of the engendering of a theory or experiment or technique of judgement or whatever, it would make redundant the normative problem of whether on that occasion the appropriate task-related standards had been maintained. Should a theory have been accepted on that evidence? Should the practice of accepting or rejecting theories on relevant evidence be adhered to? If one were to essay the task of explaining socio-historically why a community had the practice of assessing theories this way, as if success in the latter precluded a philosophical explanation of the attractions of the practice, one would be committing the genetic fallacy at a second order, so to speak.

I conclude that while there is supposed to be a causal relation between social factors and knowledge the logical conditions of independent description of cause and of effect rules out the reduction of knowledge to a social 'entity' unless some further, independent argument is advanced. And while knowledge can be defined independently of its social engendering conditions any attempt to exclude philosophical accounts of knowledge or of scientific practices that are supposed to lead to it are commissions of the genetic fallacy. By cutting out the causal relation the Paris school offer a more radical sociology of knowledge.

The Paris Programme

In the analyses of Latour and Woolgar[16], Knorr-Cetina (*op. cit.,*) and others, knowledge and truth are not conserved. They are fully redefined as social phenomena. The Paris school do not propose a causal relation between a social fact external to the scientific community and that community's beliefs and decisions. 'Knowledge' and 'truth' are identical with certain social attributes of the community itself. Essentially the Paris school offer us an anthropological study of a tribe. Their view implies that scientists are not a community in the ordinary sense. They are more like a tribe—a folk—the 'science-folk'. I will borrow a word from 'generalized Scandinavian' and call them the 'Videnskap-folk'.

a. Science as 'writing'

In one sense the Paris school is not Marxist at all—that is, they do not relate the structure of the scientific community to. the economic order of society and its development to a classical class dialect. Nevertheless they use Marxist concepts in their analysis of the social structure of the Videnskap-folk. Marxist theory provides an analogue in that they treat the scientific community as analogous to a capitalist society. But the Videnskap-folk are autonomous from their surrounding society whether it is capitalist, socialist, or what. So their analysis should apply to an institute in the Moscow Academy of Sciences or to a laboratory in California, or anywhere else. The philosophical theories used in the analysis of the tribe's activities derive from the ideas of Derrida and of Bourdieu. The central insight comes from Derrida that the scientific community is a folk or tribe, whose main business is the production of *writing*. The structure of the tribe is dependent upon the form of the writing, so that 'texte'-structure determines social structure. I will be analyzing the social structure independently, but only to show in the end how it is determined by the exigencies of the product. 'Texte' is the product and the form of text determines the social structure for producing it. With this as the basic insight the Paris school seem to be claiming that nothing that scientists do cannot be explained by references to the anthropology of the Videnskap-folk. The first step in analyzing the culture of this society is to bracket the meaning of the product, that is to ignore, what the scientific theory is actually saying. By bracketing content one can look only at the practices involved in the production of the writing as a product.

b. *The Videnskap-folk as a class-society*

Any scientific institution, it might be a university department, a laboratory in a chemical company, or an Institute in the Soviet Academy of Sciences, etc. consists of three social classes. We can distinguish these classes both empirically and theoretically. In the upper class are people with titles like 'Professor', in the middle class are lecturers, senior research officers, and so on, while in the lower class we find technicians, graduate students and the like. We are familiar with the terminology of academic ranks, but in the analysis of the Paris school, these become social classes, upper, middle and lower. The traditional epistemological categories, for instance those we are accustomed to call 'knowledge', and 'truth' and 'falsity' etc. are engendered by dialectical relations between the classes. There is a lower/middle dialectic in which categories like 'truth' and 'falsity' are engendered, and there is a middle/ upper dialectic in the course of which we find something produced which we have traditionally called knowledge. Empirically the classes can be distinguished in various ways, for instance by assymetrical acts of deference, by power relations, and so on; they have many of the characteristics of traditional social classes. The amount of personal contact is another indicator, regulated perhaps by the sanctity of personal territory. In one institute in the study, the Director was not able physically to make contact with the people further down in the system. He was divided off by a glass wall. The only way to communicate with him was to *write* something. As in traditional social classes, clothes are used as class-markers. The upper class wear ties, the middle class are dressed more informally, while the lower class wear the white coats of technicians, to show perhaps how uncontaminated by impurities are their analyses. We are beginning an anthropological investigation.

A theoretical distinction emerges by identifying distinguishable roles in the production of the typical product of the Videnskap-folk, scientific writing. The classes perform quite different tasks in the total work involved in writing a scientific paper. But we will be able to distinguish the classes theoretically only when we have analyzed the product and identified the necessary steps in its production.

Let us now turn to the product, the writings. How do we know that scientific institutions produce writing? Commonsense tells us that scientific institutions produce *knowledge*, certified factual truth. But if one takes the anthropological stance and watches what comes in and what goes out of an institution manned by members of the Videnskap-folk it is overwhelmingly writing— books, offprints, pre-prints, referees' reports, grant requests, testimonials and so on. The obvious conclusion is that the institution is there to produce writing. Let us look at the nature of the writing. What sort of writing operations are there? In a scientific institution there are two writing devices. There are laboratories and there are desks. In a laboratory there is apparatus. The lower class, graduate students and technicians, are working with this apparatus; it might include, for example, a cage of experimental animals, say twelve rats. The technicians feed the rats various diets specified by the research workers of the middle class. But when the results of the experiments are transferred from the lower class to the middle class the technicians do not bring the experiment itself. The rats are turned into writing. If three rats are dead, they are not carried by their tails and dumped on the research officer's desk. A technician might see a long piece of paper covered in holes and dots, or in the case in point you might pass up the *sentence*, '75 per cent of the rats survived treatment A'. From the point of view of the anthropologist accounting for the behaviour of the Videnskap-folk the fact that the experiment has a relationship with the world is relevant only in so far as it engenders writing.

What is happening to that writing? It appears on a desk. For the first time in philosophy Latour and Woolgar offer us a general theory of the desk. The desk has a characteristic layout. Usually on the left-hand side can be found a clutter of very strange things. Airline tickets, with writing on their backs; bits of print cut out of old typescript; scraps of paper glued or stapled together, snippets of xeroxes. This is the creative corner. On the other side is a printed book, article or paper, taken from the archives of the tribe (the 'library' in our naive terminology). This is part of the scriptures. It has authoritative point. In the middle of the desk is a pad on which new writing is created by combining writings from left- and right-hand sides of the desk. But this writing can be further analyzed. It is class-related. Some of the writing on the left-hand side comes from the creative powers of the middle-class writer; some comes up from the lower classes. Scripture, as the writings of authenticated and reputable scientists is upper class. The middle class writer has the job of combining these writings of upper and lower class origins into new writings. How do the upper class create scriptures? Middle class writing, or 'draft articles' as we call them, are sent up to the local representative of the upper class—the Director. What does he do with it? The Director's act of writing is to put his name to the work along with the names of the middle-class persons involved. The lower-class person usually does not get his name on as part author. The third activity is signing. These writing activities are necessitated by the labour requirement of manufacturing a product, 'knowledge'. According to the theses of the Paris school, we have now described all the relevant activities of an institution, since the aim of the institution is to produce signed writing. But what is the point of all this activity? What is the enterprise for?

We can understand that only in terms of the dialectics of class interests. In these are produced what we naively call 'truth' and 'knowledge'. Why do we produce this? The theory of interest derives from an analogue of Marxist analysis. There are three classes. They have divergent interests and the dialectic between them is powered by those divergent interests. What are these interests? So far we have been talking, I think, in the mode of Derrida. Now we must start to talk in the mode of Bourdieu. What are scientific institutions really for? The anthropological stance shows that scientific institutions are really for creating symbolic capital, that is reputation. The enterprise of producing the writings, signing them and so on, is to create a structure in which the upper class can accumulate symbolic capital. The Marxist analysis can be taken a step further. The dialectic of the class system of the Videnskap-folk is 'the logic of capital'; in short, symbolic capital is being used to create more symbolic capital. The Director's scientific reputation is used to expand the scope of his institution, to obtain the wherewithal to hire more workers, and so on. Symbolic capital works exactly like monetary capital. It is in the business of producing more of itself just like money capital. The fact that the capitalist system produces cars is incidental in a Marxist analysis, because the logic of capital is to produce more capital, not specifically to produce cars. If in the scientific community written papers are produced, that is only of importance in creating surplus value.

Interests enter into the relationship between the lower and the middle as a dialectic of trust. The reputation of a worker in a laboratory, and technician etc., is determined by the extent to which the people for whom he is working trust his results. He has an interest in making the apparatus work. The dialectic between the lower and the middle arises because the distrust or trust which powers it has to do with the maintenance of the belief that the apparatus has worked and worked according to specification. What counts as 'working' for the apparatus has been specified by the middle class—whether the apparatus meets the prescription shows in the work of the lower class. The dialectic between the middle and the upper is powered by an interest in symbolic capital, in using each other to create reputation. The upper need the middle to do the work, but the middle are essentially in competition with the upper for the capital. The person in the middle wants to put his name next to the director to get the work published, and so he will begin to accumulate symbolic capital for himself, be appointed to a professorship in Bergen, then Oslo, and eventually his reputation will rise so high that he becomes Vice-Rector and can hardly find time for any more scientific work. By supporting the 'stars' one can become a star oneself. As the star declines his former colleagues desert him.

This is an outline of the anthropological analysis. How does truth and knowledge come into it? The idea of the *apparatus working* is central to the analysis of truth. Truth enters into the system in so far as the pieces of apparatus work, according to the specifications of the makers and the hypotheses and predictions of the middle class. Arguments between the technicians and researchers, the people working in the system, will determine what truth is, relative to a given apparatus. Similarly, knowledge will be the result of the dialectic of reputation. By virtue of the stars, refereeing and publishing a writing, it enters the scriptures and counts as knowledge. It will become the source of true beliefs for the purposes of that scientific community.

The 'knowledge' and 'truth' defined by these class-dialects is relative to each particular institution. So closely is a body of beliefs tied to particular moities of the Videnskap-folk that for the Paris school it is indexical, that is meaningful only within a specific institution. For example, what is the accepted result of an experiment, depends partly on the apparatus and technicians possessed by an institution. The concept of truth-as-working is indexical of a particular laboratory. The dependence of indexical 'truth' on technicians leads to Directors behaving like Victorian ladies stealing each other's cooks. They try to steal each other's technicians. The capitalist analysis can be further extended, with each laboratory or research institute and university department seen as analogous to a firm, at least in competing for 'symbolic' capital.

The analysis I have just sketched can be treated as anthropology, a contribution to social science in its own right, achieved by the technical device of bracketing content. As such it seems to be of very great interest. But it is also presented as epistemology. In that guise the bracketing of content has to be justified by demonstrating that there is nothing scientists do that depends on the content of their writings. I turn now to show how in this guise the work of the Paris school involves itself with various philosophical confusions and evasions of important issues which are such that it must be rejected as a contribution to epistemology.

1. What is the status of the analysis I have just presented? Is it just a contribution to my stock of symbolic capital, or does it involve the traditional epistemic concepts which we allegedly suspended by the bracketing of content? Suppose, for instance, the

photographs used by Latour and Woolgar as evidence turned out to be fakes? If we discovered that Latour was faking his photographs, then we would reject one of the premises of his theory on the grounds of it being false.

This objection can be generalized. If *all* explanatory theories are held only relative to interests, so is that statement itself. I do not share the interests in reductive sociology of either the Edinburgh or the Paris school, so relative to my interests the statement can be rejected. Unless there is some further way in which proper and improper interests can be distinguished the reductive theory, in whatever form, is self-refuting. Both the Edinburgh and Paris schools seem to imply in their writings that everyone else's interests but their own are somehow 'improper', for instance class or career related. If, as some sociologists have recently tended to do, we distinguish between non-cognitive and cognitive interests, then by declaring a cognitive interest, so to speak, I 'unbracket' the content of a theory and bring it under judgement in the traditional way.

2. Both the Edinburgh and the Paris school adopt a philosophical theory of science without argumentative defence. The Edinburgh school adopt a form of conventionalism to justify their bracketing of content. For the Paris school to identify 'truth' with the dialectic between the lower and the middle class, a form of positivistic instrumentalism is presumed. Only in that way could we identify the proper outcome of scientific work with instrumental 'working'. But this theory of science is not defended. Therefore, the work of Latour, Woolgar, Knorr-Cetina, etc. is not a pure anthropological theory.

3. In considering the dialectic we have to ask whether the relationship between the lower and middle class would permit an indefinitely open dispute over 'working' terminated only by power relations? Could one continue to dispose of dead rats indefinitely? Suppose of the twelve rats given a treatment eleven have died. A research officer passionately wedded to the theory which entailed the safety of the treatment makes those eleven dead rats disappear, once or twice, by blaming the technician. Eveunally one has to be take account of the dead rats as, in some way, a brute fact. There is a kind of final intransigence about the world which you cannot make away with. But if the research officer reasons as follows: 'My hypothesis was wrong. This stuff is poisonous to rats', he has shifted from an interest-motivated social

action to a deployment of an instrument of traditional epistemology—*modus tollens*.

4. In expounding the theory that the 'logic' of the social processes for the acquisition of symbolic capital will explain everything that the tribe does, that the empirically observed avarice in the accumulation of reputations is the only kind of theory required to understand a social order, we have described its economy, but we have not mentioned its morality. But any tribe, and the Videnskap-folk are no exception, not only have an economy, but a morality. We have to take account of the morality of the tribe if we are to understand why the tribe does certain things. Why do people who fake results get exiled from the Videnskap-folk? Why was contempt heaped upon Sir Cyril Burtt? He did not violate the economic imperatives of the Videnskap-folk. He violated their moral imperatives. In investigating a tribe anthropologically we must take account not only of their economic system, but of their moral order, and then we find not only 'career' interests but cognitive interests. I believe that the concepts of 'truth' and 'knowledge' are not part of the working epistemology of science. In this book I have referred to theories as satisfactory or unsatisfactory, not true or false. The epistemological criticisms of the strict concepts of truth and knowledge from Hume to Popper are correct, I believe. But those criticisms do nothing to show that these concepts have no place in scientific thinking as part of a morality. In practice, in science we use different epistemological concepts. These define the actual cognitive interests of scientists who, no doubt, have other interests as well. My thesis is that we cannot understand science wholly in terms of its epistemology. We have to understand it also in terms of its morality, not as we understand the society of scientists in terms of its 'economy'.

In a comprehensive theory of science we should have a place for the moral order of the scientific community as well as its epistemology and its career structure. The Paris school have left out a crucial feature of the anthropology of the tribe. No anthropologist would seriously offer an ethnography unless it included the moral system of the folk as well as their actual social practices. Truth and knowledge, I believe, are not empirical concepts, like, say 'blue', nor everyday moral concepts like 'honest'. They are ideal specifications of worth, like 'good' and 'beautiful'. Part of the trouble that we have had in trying to understand them is that we have to treat them as if they were descriptions of scientific work and its products rather than representations of the scientific morality.

5. The Paris school depends on treating the contents of theories as irrelevant to understanding why they are accepted or rejected by the Videnskap-folk. But suppose we ask why they have persisted for so long if all they do is produce writings? In short, can we really bracket the content of these writings? I want to draw attention to one feature of the content, namely, its rationality, which perhaps can explain why the tribe exists. The Paris school have a theory of rationality, according to which rationality is deliberately created in order to engage in controversies and arguments. It is a competitive advantage in the competition between firms, like good petrol consumption. If we ask, why do scientists make their scientific papers rational?, the answer is not because in this way they will thereby improve their understanding of the world, but because if scientist A detects irrationality in the writings of Scientist B, the latter will lose any argument he engages in with A. So there is a social reason for putting in rationality.

Why would the tribe have just this way of evaluating products? This question cannot be answered by bracketing content, for by so doing we have confined ourselves to a study of the vehicle, the writing, and we have eliminated from consideration what the vehicle is carrying, its 'load'. The contents of scientific papers purport to be descriptions of things and processes in the real world. If that is true we have an explanation of why the Videnskap-folk value rationality. Otherwise their valuing that property of discourse rather than any other is a mere historical accident.

REFERENCES

1. W. Whewell, *The Philosophy of the Inductive Sciences*, Original edition, London, 1847; Johnson Reprint Corporation, New York, 1967; Book I.
2. N. R. Hanson, *The Concept of the Positron*, Cambridge University Press, 1963.
3. J. J. Gibson, *The Senses Considered as Perceptual Systems*, George Allen & Unwin, London, 1968.
4. P. K. Feyerabend, *Against Method*, New Left Books, London, 1975.
5. D. Papineau, 'The *Vis Viva* controversy: Do meanings matter?', *Studies in History and Philosophy of Science*, **8**, 1977, 111–142.
6. H. Putnam, *Philosophical Papers*, Vol. 2, Cambridge University Press, 1975.
7. S. E. Toulmin, *Human Understanding*, Clarendon Press, Oxford, 1972.

8. R. Harré, *Great Scientific Experiments,* Phaidon Press, Oxford, 1982.
9. T. S. Kuhn, *The Structure of Scientific Revolutions,* University of Chicago Press, 1962.
10. D. Shapere, 'Meaning and scientific change', in R. G. Colodny (ed.), *Mind and Cosmos,* University of Pittsburgh Press, 1966, pp. 41–85.
11. K. Knorr-Cetina, *The Manufacture of Knowledge,* Pergamon Press, Oxford, 1982.
12. J. Ravetz, *Scientific Knowledge and Its Social Problems,* Clarendon Press, Oxford, 1977.
13. D. Bloor, *Knowledge and Social Imagery,* Routledge & Kegan Paul, London, 1976.
14. B. Barnes, *T. S. Kuhn and Social Science,* Macmillan, London, 1982.
15. B. Barnes, *Interests and the Growth of Knowledge,* Routledge & Kegan Paul, Oxford, 1977.
16. B. Latour and A. Woolgar, *Laboratory Life,* Sage, Los Angeles, 1979.

Further Reading

INTRODUCTORY

J. T. Davies, *The Scientific Approach,* Academic Press, London and New York, 1965.

K. Lambert and G. D. Brittan, *An Introduction to the Philosophy of Science,* Prentice-Hall, Englewood Cliffs, NJ, 1970.

J. Losee, *An Historical Introduction to the Philosophy of Science,* Oxford University Press, London, Oxford and New York, 1972.

S. Toulmin, *The Philosophy of Science,* Hutchinson, London, 1953 (subsequent reprints).

ADVANCED

1. *Concerning Scientific Description*
   B. Ellis, *Basic Concepts of Measurement,* Cambridge University Press, 1968
   N. R. Hanson, *Patterns of Discovery,* Cambridge University Press, 1958.
   D. Papineau, *Theory and Meaning,* Clarendon Press, Oxford, 1979.
2. *Concerning Scientific Explanation*
   M. Bunge, *Method, Model and Matter,* Reidel, Dordrecht, 1973.
   C. G. Hempel, *Aspects of Scientific Explanation,* Free Press, New York, 1965.
   M. B. Hesse, *The Structure of Scientific Inference,* Macmillan, London, 1974.
3. *Concerning Inductive and Non-Inductive Reasoning*
   S. F. Barker, *Induction and Hypothesis,* Cornell, 1957.
   N. Rescher, *Induction,* University of Pittsburgh Press, 1980.
   E. Sober, *Simplicity,* Clarendon Press, Oxford, 1975.
4. *Concerning Rationality of Science*
   P. Feyerabend, *Against Method,* New Left Books, London, 1975.
   K. Knorr-Cetina, *The Manufacture of Knowledge,* Pergamon Press, Oxford, 1981.
   W. Newton-Smith, *The Rationality of Science,* Routledge & Kegan Paul, London, 1982.
5. *Major Works*
   E. Nagel, *The Structure of Science,* Routledge & Kegan Paul, London, 1981.
   M. Polanyi, *Personal Knowledge,* Routledge & Kegan Paul, London, 1958, etc.
   K. R. Popper, *The Logic of Scientific Discovery,* Hutchinson, London, 1959, etc.

Index of Names

Index of Topics